Size Limits of Very Small Microorganisms

Proceedings of a Workshop

Steering Group for the Workshop on Size Limits of Very Small Microorganisms
Space Studies Board
Commission on Physical Sciences, Mathematics, and Applications
National Research Council

NATIONAL ACADEMY PRESS
Washington, D.C.

NOTICE: The project that is the subject of this report was approved by the Governing Board of the National Research Council, whose members are drawn from the councils of the National Academy of Sciences, the National Academy of Engineering, and the Institute of Medicine. The members of the steering group responsible for the report were chosen for their special competences and with regard for appropriate balance.

The National Academy of Sciences is a private, nonprofit, self-perpetuating society of distinguished scholars engaged in scientific and engineering research, dedicated to the furtherance of science and technology and to their use for the general welfare. Upon the authority of the charter granted to it by the Congress in 1863, the Academy has a mandate that requires it to advise the federal government on scientific and technical matters. Dr. Bruce Alberts is president of the National Academy of Sciences.

The National Academy of Engineering was established in 1964, under the charter of the National Academy of Sciences, as a parallel organization of outstanding engineers. It is autonomous in its administration and in the selection of its members, sharing with the National Academy of Sciences the responsibility for advising the federal government. The National Academy of Engineering also sponsors engineering programs aimed at meeting national needs, encourages education and research, and recognizes the superior achievements of engineers. Dr. William A. Wulf is president of the National Academy of Engineering.

The Institute of Medicine was established in 1970 by the National Academy of Sciences to secure the services of eminent members of appropriate professions in the examination of policy matters pertaining to the health of the public. The Institute acts under the responsibility given to the National Academy of Sciences by its congressional charter to be an adviser to the federal government and, upon its own initiative, to identify issues of medical care, research, and education. Dr. Kenneth I. Shine is president of the Institute of Medicine.

The National Research Council was organized by the National Academy of Sciences in 1916 to associate the broad community of science and technology with the Academy's purposes of furthering knowledge and advising the federal government. Functioning in accordance with general policies determined by the Academy, the Council has become the principal operating agency of both the National Academy of Sciences and the National Academy of Engineering in providing services to the government, the public, and the scientific and engineering communities. The Council is administered jointly by both Academies and the Institute of Medicine. Dr. Bruce Alberts and Dr. William A. Wulf are chairman and vice chairman, respectively, of the National Research Council.

Support for this project was provided by Contract NASW 96013 between the National Academy of Sciences and the National Aeronautics and Space Administration. Any opinions, findings, conclusions, or recommendations expressed in this material are those of the authors and do not necessarily reflect the views of the sponsor.

International Standard Book Number 0-309-06634-4
Copyright 1999 by the National Academy of Sciences. All rights reserved.

COVER: Design by Penny Margolskee.
Copies of this report are available free of charge from:

Space Studies Board
National Research Council
2101 Constitution Avenue, NW
Washington, DC 20418

Printed in the United States of America

STEERING GROUP FOR THE WORKSHOP ON SIZE LIMITS OF VERY SMALL MICROORGANISMS

ANDREW KNOLL, Harvard University, *Co-chair*
MARY JANE OSBORN, University of Connecticut Health Center, *Co-chair*
JOHN BAROSS, University of Washington
HOWARD C. BERG, Harvard University
NORMAN R. PACE, University of California at Berkeley
MITCHELL SOGIN, Marine Biological Laboratory

Staff

SANDRA J. GRAHAM, Study Director (from October 17, 1998)
JOSEPH L. ZELIBOR, Jr., Study Director (through October 16, 1998)
ERIN C. HATCH, Research Associate
JACQUELINE D. ALLEN, Senior Program Assistant (through February 1999)
THERESA M. FISHER, Senior Program Assistant (from April 1999)
LAURA OST, Consultant

SPACE STUDIES BOARD

CLAUDE R. CANIZARES, Massachusetts Institute of Technology, *Chair*
MARK R. ABBOTT, Oregon State University
FRAN BAGENAL, University of Colorado
DANIEL N. BAKER, University of Colorado
LAWRENCE BOGORAD, Harvard University*
DONALD E. BROWNLEE, University of Washington*
ROBERT E. CLELAND, University of Washington
GERARD W. ELVERUM, JR., TRW Space and Technology Group
ANTHONY W. ENGLAND, University of Michigan*
MARILYN L. FOGEL, Carnegie Institution of Washington
RONALD GREELEY, Arizona State University*
BILL GREEN, former member, U.S. House of Representatives
JOHN H. HOPPS, JR., Morehouse College
CHRIS J. JOHANNSEN, Purdue University
ANDREW H. KNOLL, Harvard University
RICHARD G. KRON, University of Chicago
JONATHAN I. LUNINE, University of Arizona
ROBERTA BALSTAD MILLER, Columbia University
BERRIEN MOORE III, University of New Hampshire*
GARY J. OLSEN, University of Illinois at Urbana-Champaign
MARY JANE OSBORN, University of Connecticut Health Center
SIMON OSTRACH, Case Western Reserve University*
MORTON B. PANISH, AT&T Bell Laboratories (ret.)*
CARLÉ M. PIETERS, Brown University*
THOMAS A. PRINCE, California Institute of Technology
PEDRO L. RUSTAN, JR., U.S. Air Force (ret.)
JOHN A. SIMPSON, University of Chicago*
GEORGE L. SISCOE, Boston University
EUGENE B. SKOLNIKOFF, Massachusetts Institute of Technology
EDWARD M. STOLPER, California Institute of Technology*
NORMAN E. THAGARD, Florida State University
ALAN M. TITLE, Lockheed Martin Advanced Technology Center
RAYMOND VISKANTA, Purdue University
PETER W. VOORHEES, Northwestern University
ROBERT E. WILLIAMS, Space Telescope Science Institute*
JOHN A. WOOD, Harvard-Smithsonian Center for Astrophysics

JOSEPH K. ALEXANDER, Director

---
*Former member.

COMMISSION ON PHYSICAL SCIENCES, MATHEMATICS, AND APPLICATIONS

PETER M. BANKS, ERIM International, Inc., *Co-chair*
W. CARL LINEBERGER, University of Colorado, *Co-chair*
WILLIAM BROWDER, Princeton University
LAWRENCE D. BROWN, University of Pennsylvania
MARSHALL H. COHEN, California Institute of Technology
RONALD G. DOUGLAS, Texas A&M University
JOHN E. ESTES, University of California at Santa Barbara
JERRY P. GOLLUB, Haverford College
MARTHA P. HAYNES, Cornell University
JOHN L. HENNESSY, Stanford University
CAROL M. JANTZEN, Westinghouse Savannah River Company
PAUL G. KAMINSKI, Technovation, Inc.
KENNETH H. KELLER, University of Minnesota
MARGARET G. KIVELSON, University of California at Los Angeles
DANIEL KLEPPNER, Massachusetts Institute of Technology
JOHN KREICK, Sanders, a Lockheed Martin Company
MARSHA I. LESTER, University of Pennsylvania
M. ELISABETH PATÉ-CORNELL, Stanford University
NICHOLAS P. SAMIOS, Brookhaven National Laboratory
CHANG-LIN TIEN, University of California at Berkeley

NORMAN METZGER, Executive Director

Foreword

The world was galvanized in August 1996 by the announcement of possible evidence for relic biogenic activity in the martian meteorite ALH84001. Prominent among the five features cited in support of this startling hypothesis was the observation of "carbonate globules and features resembling terrestrial microorganisms, terrestrial biogenic carbonate structures, or microfossils."[1] The structures, revealed in electron micrographs, range in length from 10 to 200 nm. One reason for skepticism about the claim that these have biologic origin is that the martian structures are generally much smaller than the terrestrial objects to which they were compared.

Regardless of one's conclusions about the meteoric evidence for life on Mars, the public fanfare was very effective in focusing attention on scientific questions central to understanding if and how we can recognize extraterrestrial life. The topic of the workshop whose findings are reported here, the size limits of very small microorganisms, is an important one for the interpretation of the carbonate structures on ALH84001, as preparation for future investigations of other samples carried to Earth naturally or in spacecraft and as a litmus test of how well we understand biological organization in general. An indicator of the interest and excitement catalyzed by the Mars announcement is that nearly all the panelists invited to participate accepted with alacrity.

Nearly two dozen researchers applied their diverse expertise to the problem of extrapolating from what we know about Earth's abundant microbial population and the laws of physics and chemistry to draw conclusions about size limits for putative extraterrestrial life forms. Extrapolation is necessary because the only thing we can be reasonably confident about is that extraterrestrial organisms will differ in significant ways from those we find around us. Our life forms have been described by a Nobel Prize-winning biologist as Rube Goldberg contraptions assembled over eons by the stochastic processes of evolution—one can hardly expect that these would be exactly reproduced in other environments.

Part of the legacy of the ALH84001 meteorite is a significant increase in the vigor of NASA's programs in astrobiology, the exploration of the context and possible evidence for life elsewhere in the

---

[1] D.S. McKay, et al. (1996). *Science* **273**:924-930.

solar system, and the search for other planetary systems that might harbor life. These are pursuits that stimulate scientists and the public alike. They also demand the highest standards of scientific rigor—as stated by the late Carl Sagan, extraordinary claims demand extraordinary evidence. This volume lives up to those standards.

Claude R. Canizares
Chair, Space Studies Board

Preface

The question of minimal microbial size continues to be a subject of debate within the scientific community. There is no widely accepted theoretical minimum size for microorganisms. In examining samples from Mars and elsewhere for signs of life, scientists need to know what to look for and how to interpret the results. To help guide its activities in this area, NASA's Office of Space Science (OSS) requested that the Space Studies Board organize a workshop to provide a forum for discussions of the theoretical minimum size for microorganisms. The Board formed the Steering Group for the Workshop on Size Limits of Very Small Microorganisms, which organized a workshop of leading experts in fields relevant to this question.

The researchers who participated in the two-day workshop, convened on October 22-23, 1998, at the facilities of the National Academy of Sciences in Washington, D.C., addressed the following questions:

1. What features of biology characterize microorganisms at or near nanometer scale? Is there a theoretical size limit below which free-living organisms cannot be viable? If we relax the requirement that cells have the biochemical complexity of modern cells, can we model primordial cells well enough to estimate their likely sizes?

2. Is there a relationship between minimum cell size and environment? Is there a continuum of size and complexity that links conventional bacteria to viruses? What is the phylogenetic distribution of very small bacteria?

3. Can we understand the processes of fossilization and non-biological processes sufficiently well to differentiate fossils from artifacts in an extraterrestrial rock sample?

4. Does our current understanding of the processes that led from chemical to biological evolution place constraints on the size of early organisms? If size is not constrained, are there chemical signatures that might record the transition to living systems?

Workshop participants were organized into four panels. Each panel was coordinated by a moderator, who provided a brief introduction to the panel topic and set the stage for the panel discussion. The moderator ensured that the panelists debated issues in a constructive and scientific manner and that all sides of the issues were explored. Prior to the workshop, each panelist submitted a short paper with a critical assessment of issues. Each panelist made a presentation during the workshop and was later given the opportunity to add to his or her paper any additional points made during that presentation. The content and views expressed in these papers are solely the responsibility of the individual authors.

Although the workshop did not provide definitive answers to the questions addressed, this proceedings document—which describes the workshop findings and identifies issues and opportunities for future research areas to improve our understanding of the size limits of microorganisms—offers novel perspectives and insights, and it provides an intellectual framework for further exploration of key issues discussed by workshop participants.

Comprising eighteen invited papers and a summary of each of the four panel discussions, this volume points out areas in which participants reached general consensus. It does not offer conclusions and recommendations. It is anticipated that this material will provide a valuable reference for astrobiology research and efforts related to the examination of samples returned from Mars and planetary satellites.

Acknowledgment of Reviewers

This report has been reviewed by individuals chosen for their diverse perspectives and technical expertise, in accordance with procedures approved by the National Research Council's (NRC's) Report Review Committee. The purpose of this independent review is to provide candid and critical comments that will assist the authors and the NRC in making the published report as sound as possible and to ensure that the report meets institutional standards for objectivity, evidence, and responsiveness to the study charge. The contents of the review comments and draft manuscript remain confidential to protect the integrity of the deliberative process. We wish to thank the following individuals for their participation in the review of this report:

Jack Farmer, Arizona State University,
Marilyn L. Fogel, Carnegie Institution of Washington,
Jeffrey Lawrence, University of Pittsburgh,
Marsha I. Lester, University of Pennsylvania, and
Gary J. Olsen, University of Illinois at Urbana-Champaign.

Although the individuals listed above have provided many constructive comments and suggestions, responsibility for the final content of this report rests with the the individual contributing authors, the workshop steering group, and the NRC.

Contents

OVERVIEW (Andrew Knoll, Steering Group Co-chair)   1

PANEL 1   5
    Discussion (Summarized by Christian de Duve, Panel Moderator, and Mary Jane Osborn, Steering Group Co-chair), 5
    Metabolism and Physiology of Conventional Bacteria (Dan G. Fraenkel), 10
    A Biophysical Chemist's Thoughts on Cell Size (Peter B. Moore), 16
    Correlates of Smallest Sizes for Microorganisms (Monica Riley), 21
    Mechanical Characteristics of Very Small Cells (David Boal), 26
    Gene Transfer and Minimal Genome Size (Jeffrey G. Lawrence), 32

PANEL 2   39
    Discussion (Summarized by Kenneth Nealson, Panel Moderator), 39
    Can Large dsDNA-Containing Viruses Provide Information about the Minimal Genome Size Required to Support Life? (James L. Van Etten), 43
    Suggestions from Observations on Nanobacteria Isolated from Blood (E. Olavi Kajander, Mikael Björklund, and Neva Çiftçioglu), 50
    Properties of Small Free-Living Aquatic Bacteria (D.K. Button and Betsy Robertson), 56
    Bacteria, Their Smallest Representatives and Subcellular Structures, and the Purported Precambrian Fossil "Metallogenium" (James T. Staley), 62
    Smallest Cell Sizes Within Hyperthermophilic Archaea ("Archaebacteria") (Karl O. Stetter), 68
    The Influence of Environment and Metabolic Capacity on the Size of a Microorganism (Michael W.W. Adams), 74
    Diminutive Cells in the Ocean—Unanswered Questions (Edward F. DeLong), 81

PANEL 3 85

Discussion (Summarized by Andrew Knoll, Panel Moderator), 85
Fossils and Pseudofossils: Lessons from the Hunt for Early Life on Earth
    (J. William Schopf), 88
Taphonomic Modes in Microbial Fossilization (Jack Farmer), 94
Investigation of Biomineralization at the Nanometer Scale by Using Electron
    Microscopy (John Bradley), 103

PANEL 4 107

Discussion (Summarized by Leslie Orgel, Panel Moderator, and
    Laura Ost, Consultant), 107
Primitive Life: Origin, Size, and Signature (James P. Ferris), 111
Constraints on the Sizes of the Earliest Cells (Jack W. Szostak), 120
How Small Can a Microorganism Be? (Steven A. Benner), 126

APPENDIXES
    A Steering Group Biographies, 139
    B Request from NASA, 142
    C Workshop Agenda, 143
    D Workshop Participants, 147

This report is dedicated to the memory of
Dr. Joseph L. Zelibor, Jr.
(1953-1998)
*a respected friend and colleague
whose boundless energy and vision shaped this workshop.*

Overview

Andrew Knoll, Steering Group Co-chair

How small can a free-living organism be? On the surface, this question is straightforward—in principle, the smallest cells can be identified and measured. But understanding what factors determine this lower limit, and addressing the host of other questions that follow on from this knowledge, require a fundamental understanding of the chemistry and ecology of cellular life. The recent report of evidence for life in a martian meteorite and the prospect of searching for biological signatures in intelligently chosen samples from Mars and elsewhere bring a new immediacy to such questions. How do we recognize the morphological or chemical remnants of life in rocks deposited 4 billion years ago on another planet? Are the empirical limits on cell size identified by observation on Earth applicable to life wherever it may occur, or is minimum size a function of the particular chemistry of an individual planetary surface?

These questions formed the focus of a workshop on the size limits of very small organisms, organized by the Steering Group for the Workshop on Size Limits of Very Small Microorganisms and held on October 22 and 23, 1998. Eighteen invited panelists, representing fields ranging from cell biology and molecular genetics to paleontology and mineralogy, joined with an almost equal number of other participants in a wide-ranging exploration of minimum cell size and the challenge of interpreting micro- and nano-scale features of sedimentary rocks found on Earth or elsewhere in the solar system. This document contains the proceedings of that workshop. It includes position papers presented by the individual panelists, arranged by panel, along with a summary, for each of the four sessions, of extensive roundtable discussions that involved the panelists as well as other workshop participants.

CONSENSUS AND CAVEATS

The discussions forming the basis of this document sought to address three distinct but related issues: (1) What are the theoretical, observable, and empirically testable limits on the minimum size of organisms living on Earth today? (2) What, in theory, are the size limits on organisms not constrained by the biochemistry of extant cells? and (3) How can we recognize traces of ancient and potentially unfamiliar life in samples from other bodies in the solar system? As is evident from the summaries, there was strong consensus on the first issue, but the others remain open. The six geneticists and cell biologists in Panel 1 reached consensus on the smallest size likely to be attained by organisms of modern biochemical complexity. Free-living organisms require a minimum of 250 to 450 proteins along with the genes and ribosomes necessary for their synthesis. A sphere capable of holding this minimal molecular complement would be 250 to 300 nm in diameter,[1] including its bounding membrane. Given the uncertainties inherent in this estimate, the panel agreed that 250 ± 50 nm constitutes a reasonable lower size limit for life as we know it. At this minute size, membranes have sufficient biophysical integrity to contain interior structures without the need for a cell wall, but only if the organism is spherical and has an osmotic pressure not much above that of its environment.

Panel 2 consisted of microbial ecologists asked to elucidate the smallest sizes actually observed in free-living organisms. Once again, consensus emerged from the panel's discussion. Consistent with the theoretical limits articulated by Panel 1, members of Panel 2 reported that bacteria with a diameter of 300 to 500 nm are common in oligotrophic environments, but that smaller cells are not. Nanobacteria[2] reported from human and cow blood fall near the lower size limit suggested by cell biologists; however, the much smaller (ca. 50 nm) bodies found in association with these cells may not, themselves, be viable organisms. Observations on archaea indicate that, in general, they have size limits similar to those for bacteria.

Two problems constrain discussions of minimal cell size in natural environments. Commonly used methods of measuring cell size have inherent uncertainties or possibilities of error. Perhaps more important, most cells found in nature cannot be cultivated. Thus, ignorance about biological diversity at small sizes remains large. These problems notwithstanding, it appears that very small size in modern organisms is an adaptation for specific environmental circumstances, including stress and scarcity of resources. Primordial organisms may or may not have been tiny, but the smallest organisms known today reside on relatively late branches of the RNA phylogeny.

Whereas Panels 1 and 2 indicated that a cell operating by known molecular rules—with DNA or maybe RNA, ribosomes, protein catalysts, and other conventional cell machinery—would have a lower size limit of 200 to 300 nm in diameter, Panel 4 suggested that primitive microorganisms based on a single-polymer system could be as small as a sphere 50 nm in diameter. There is no assurance that primordial cells would have been this small or, if they were, that such minute cells would have been more than transitory features of early evolution. Nonetheless, unless one is willing to posit that everywhere it has arisen, life has evolved a biochemical machinery comparable to that seen on Earth, the rules that govern minimum cell size may not be universal.

In fact, as explored by Panel 3, there are a number of ways that living cells or fossils might fall below the minimum size deemed likely by cell biologists and ecologists. On Mars or Europa, fossils might preserve a record of biological systems different from those we understand—perhaps early products of evolution that made do with a small complement of functional molecules. Organisms of modern biochemistry might become small by being pathogens or living in consortia—that is, by using the products of another organism's genes. Or, fossils might preserve remains that shrank after death, or parts of organisms rather than complete cells—both are common in the terrestrial record.

---

[1]Contributors to the workshop have usually described relevant scale sizes or dimensions in units of nanometers (nm) or micrometers (μm), depending on the context and the features being described. For an indication of the range of relevant scale sizes, see Figure 1 in the paper by Jack Farmer, Panel 3, p. 94.

[2]While biologists have yet to agree on a precise meaning for this term, it is generally used to refer to any single-celled microorganism proposed to have a maximum diameter in the range of tens to a few hundreds of nanometers.

Of course, fossil morphologies are but one of several types of biological signature preserved in rocks. Experience with ancient terrestrial rocks shows that extractable organic molecules, minerals, fractionation in isotopic or elemental abundances, and distinctively laminated sedimentary structures can all provide indications of past life. Many of these features, however, can be mimicked by physical processes. Panel 3 concluded that a much better understanding of biological pattern formation is needed before intelligently chosen martian samples are returned to Earth. The panel also emphasized that this must go hand in hand with improved knowledge of the limits of morphological and chemical pattern formation by non-biological processes. Indigenous features of extraterrestrial samples can be accepted as biogenic only if they are incompatible with formation by physical processes.

THINKING ABOUT THE FUTURE

In 2008, a small (<1 kg) sample of martian rock and soil is scheduled to be delivered to Earth by a robotic spacecraft that will be launched to Mars in 2005. Among the important questions that will be asked of these samples is, Has Mars ever been a biological planet? Our ability to address this question is directly related to our understanding of the range of morphological features that can be produced by life and by physical processes, as well as the ranges of organic chemicals, mineral forms, and sedimentary rock features that can be generated by biological and by nonbiological processes. As the results of the workshop made clear, welcome consensus has emerged among the participants regarding the size and chemical limits on modern life on Earth. But, given reasonable uncertainty about whether such features are particular products of terrestrial evolution or universal features of life, the meter stick by which the biogenicity of martian or other planetary samples is measured will likely be knowledge of the limits on physical processes—knowledge that needs to be developed before samples from Mars arrive in the laboratory.

Panel 1

What features of biology characterize microorganisms at or near nanometer scale?

Is there a theoretical size limit below which free-living organisms cannot be viable?

If we relax the requirement that cells have the biochemical complexity of modern cells, can we model primordial cells well enough to estimate their likely sizes?

DISCUSSION

*Summarized by Christian de Duve, Panel Moderator, and
Mary Jane Osborn, Steering Group Co-chair*

Constraints on Size of a "Minimal Free-living Cell"

Constraints on the lower limits of the size of a free-living prokaryote with conventional biochemistry might be imposed by a variety of factors, including the number of protein and RNA species required for minimal essential functions; the size of the genome required to encode these essential macromolecules; the number of ribosomes necessary for adequate expression of this genome; and physical constraints, such as DNA packing or the minimum radius of curvature required for stability of a lipid bilayer membrane.

The Panel 1 moderator, Dr. de Duve, opened the workshop by commenting on some theoretical calculations of the lower limits of cell size based on the assumptions and calculations shown in Box 1 (see pp. 8-9). The representative results (Tables 1 and 2)—which are based on the unlikely assumptions that only 100 nonribosomal protein species are present in a cell, that each is present in only 10 copies, and that there is only 1 ribosome, 1 tRNA set, and 1 mRNA for each protein species—must be considered unrealistically low. Even under these stringent assumptions, such a cell would have a diameter of 206 nm, including the membrane and wall. With a more realistic assumption of 300 essential non-ribosomal protein species, the diameter would be 262 nm. If each protein were present in 1,000 copies, the diameter would be 231 nm for 100 protein species and 303 nm for 300 species (see Table 2). Thus, the minimum diameter of a spherical cell compatible with a system of genome expression and a biochemistry of contemporary character would appear to lie somewhere between 200 and 300 nm, probably closer to the latter.

Minimal Number of Essential Genes, and Impact of This Number on Minimum Cell Size

The minimal number of genes required by a saprophytic microbe living in a nutrient-rich environment was estimated from minimal requirements for metabolism, genome expression, and other essential cellular functions (Panel 1 presenters Fraenkel, Riley); comparisons of sequenced genomes (Lawrence); and the genetic capacity of the smallest known genome, that of *Mycoplasma genitalium* (Moore, Lawrence). There was general agreement among panelists and discussants that approximately 250 to 450 genes compose the set of minimal essential genes. This is strikingly consistent with the known composition of the genome of *M. genitalium*, which contains approximately 470 genes, not all of which are essential. On the other hand *M. genitalium* is an obligate parasite, lacking functions required for independent, saprophytic life. Thus, the upper limit of 450 genes is likely to be conservative.

In the discussion period, questions were raised about possible ways to reduce the genome size even further. Dr. Riley asked whether a cell wall was essential. The role of some kind of relatively rigid cell wall in preventing osmotic lysis and in maintaining cell shape was cited. Respondents noted that mycoplasmas (and bacterial L-forms) lack a rigid cell wall, but that these cells are pleomorphic and are sensitive to osmotic lysis to a greater or lesser degree. However, a cell wall is clearly not essential for cell division, and "naked" cells should be able to exist in an osmotically protective environment. However, as Dr. Ferris commented, if an organism is not spherical (or pleomorphic), it must have some kind of wall or skeletal structure to confer and maintain a defined shape (e.g., rod, spiral).

Drs. Osborn and Fraenkel wondered whether a reduction in the number of ribosomal proteins might be possible, not only eliminating additional genes, but also yielding a significantly smaller, "minimal" ribosome. However, as discussed further below, this was deemed unlikely by panelists.

Dr. de Duve estimated the contribution of the genome to the dry weight of a cell the size of *E. coli* to be on the order of 4 to 6%. These modest values, however, are almost certain to be underestimated. They should be almost doubled if only one of the two DNA strands is taken to be coding and must be increased further to the extent that the genome contains non-coding DNA. Thus, values of 10 to 15% of dry weight would seem to be acceptable for the *E. coli* genome.

Dr. Moore's presentation emphasized that, as cell volume decreases, the fraction of volume occupied by the genome increases greatly, and eventually becomes a major determinant of minimal cell size. Calculated as a fraction of cell volume, the *E. coli* genome represents a negligible contribution (0.013 g/ml, ca. 1%). However, that fraction rises to nearly 10% of cell volume (0.10 g/ml) in *M. genitalium*. Dr. Osborn asked at what density of DNA packing the transcription and replication machinery can no longer function. Dr. Moore responded that T4 phage DNA, which is functionally inert, occupies 65% of the available volume. A question was raised as to whether a single-stranded RNA genome might occupy a relatively smaller volume; however, the sense of the panelists was that the complex three-dimensional structures formed by intramolecular base-pairing would not be likely to offer significant advantage in this regard.

Constraints on Minimal Cell Size Imposed by Number and Size of Ribosomes

Dr. de Duve initially emphasized the importance of the ribosome as a major determinant of minimal cell size, noting that even a single ribosome, if surrounded by membrane and wall, would occupy a sphere of 50 to 60 nm in diameter. Dr. Riley noted that, although *E. coli*, growing in rich medium, has some 30,000 ribosomes per cell, the number of ribosomes is highly dependent on growth rate. Thus, if one allows the "minimal cell" to have a very long doubling time, the necessary number of ribosomes can be greatly reduced. As to whether a significantly smaller ribosome, containing a reduced number of protein species, might be feasible, Dr. Moore suggested that the number of proteins might be reduced by a factor of two or three, but that the resulting structure would be appallingly sloppy and inefficient by modern standards. Deletion studies in *E. coli* some 15 years ago showed that elimination of almost any of the ribosomal proteins resulted in some abnormality in the particles. Most ribosomal proteins optimize or enable the assembly of rRNA into the proper three-dimensional fold.

A Single-Polymer Model of a Minimal Primordial "Cell"

Dr. Lawrence's intriguing model, which allowed for very tiny self-replicating "cells" based on sequential horizontal transfer of single genes "wandering" among a consortium, engendered considerable interest and discussion. Questions centered on issues of evolutionary stability, scaling with an increased number of genes, and "cell" size. Dr. Szostak asked whether the model could operate as an evolutionarily stable strategy. Dr. Lawrence agreed that in any one consortium, only one cell replicates and that the system will collapse if a certain gene is lost or mutated. Dr. Orgel was concerned about how the model scales with the number of genes. Since every gene must visit each compartment ("cell") at least once, would the system work with, for example, 100 genes? Dr. Lawrence replied that replication would be very slow and would depend on how long-lived the reagents (biosynthetic intermediates) were and on the rate of gene transfer. Dr. Osborn noted that there could be great selective advantage of aggregating more than one gene into a single compartment by cell-cell fusion. Thus, even starting with the minimal mechanically stable vesicle size, the system would tend to move to larger, more efficient compartments. In summary, Dr. Lawrence emphasized that the point of the model was to illustrate that not all cells need have all the essential genetic information at the same time.

Summary and Consensus

Consensus was reached by Panel 1 participants on the following major points, assuming free-living cells with conventional biochemistry:

- A minimum of about 250 to 450 essential genes are required for viability.
- The minimal viable cell diameter is expected to lie in the range of 250 to 300 nm.
- The number of ribosomes required for adequate genome expression is a significant constraint on minimal cell size.

If, however, the requirement for coventional biochemistry and genetics is relaxed, especially with reference to primordial or exobiotic self-replicating systems, the possibility of much smaller "cells" must be considered, such as those envisioned in the single-polymer model.

Box 1
Estimating Lower Limits of Cell Size—Some Assumptions and Results

Assumptions
1. Molecular masses:
   - One DNA nucleotide = 312 Da
   - One RNA nucleotide = 324 Da
   - One amino acid residue = 110 Da
2. Both DNA strands are coding.
3. The cell contains one ribosome, one set of 20 tRNA molecules (average molecular mass 25,000 Da), and one mRNA molecule for each protein species.
4. Each ribosome consists of 50 protein molecules of average molecular mass 30,000 Da and of an equivalent quantity of RNA.
5. The cell contains N nonribosomal protein species of average molecular mass 30,000 Da, each present in 10 copies. At least 100 such protein species are deemed indispensable.
6. Wet weight = 3 × weight of (DNA + RNA + Protein).
7. Density of naked cell is 1.10.
8. Cell membrane is 6 nm thick.
9. Cell wall is 10 nm thick.

Calculations
Ribosome
RNA: 1,500,000 Da
Protein: 1,500,000 Da
$\}$ 3,000,000 Da = 5×10^{-3} fg   ($\times$ (N + 50))

Genome
   rRNAs and tRNAs: 2,000,000 Da = 3.3×10^{-3} fg
   Ribosomal proteins: 12,800,000 Da = 21×10^{-3} fg
   Other proteins: N × 255,000 Da = N × 0.42×10^{-3} fg
Total: $(24.3 + N \times 0.42) \times 10^{-3}$ fg

RNAs Other Than Ribosomal
tRNAs: 500,000 Da = 0.8×10^{-3} fg   ($\times$ (N + 50))
mRNAs: Ribosomal proteins: 13,260,000 Da = 21.9×10^{-3} fg
Other proteins: N × 265,000 Da = N × 0.44×10^{-3} fg
Total: $(61.9 + N \times 1.24) \times 10^{-3}$ fg

Other Proteins
N × 10 × 30,000 Da = N × 300,000 Da = N × 0.05×10^{-3} fg
   (DNA + RNA + Protein) = $(336.2 + N \times 6.71) \times 10^{-3}$ fg

Table 1 Cell Size

N	Dry Weight* × 10⁻³ fg	Wet Weight × 10⁻³ fg	Volume × 10⁻⁶ µm³	Diameter (nm) Naked	+ Memb.	+ Wall
100	1,007	3,022	2,747	174	186	206
200	1,678	5,035	4,577	206	218	238
300	2,349	7,048	6,407	230	242	262
450	3,356	10,067	9,152	260	272	292
950	6,711	20,132	18,302	327	339	359

* (DNA + RNA + Protein)

Table 2 Cell Composition

N	% Dry Weight (DNA + RNA + Protein)				Diameter
	Genome	Ribosomes	Other RNAs	Other Proteins	from Table 1 (nm)
Assuming each protein species present in 10 copies					
100	6.6	74.5	18.4	0.5	206
200	6.5	74.5	18.4	0.6	238
300	6.4	74.5	18.5	0.6	262
450	6.3	74.5	18.5	0.7	292
950	6.3	74.5	18.5	0.7	359
Assuming each protein species present in 1,000 copies					
100	4.4	49.9	12.4	33.3	231
200	4.1	46.8	11.6	37.5	272
300	3.9	45.7	11.3	39.1	303
450	3.8	44.8	11.1	40.3	340
950	3.7	43.8	10.9	41.6	422

As Dr. de Duve commented in opening the workshop, the results given in Tables 1 and 2 above must be regarded as unrealistically low. Referring to the values for cell composition listed in Table 2, he pointed out that in a cell with only 10 copies of each protein species, the ribosomes and the other RNA components of the protein-synthesizing machinery represent more than 90% of the dry weight. Even when 1,000 copies are present, a cell's protein-synthesizing machinery still accounts for more than 50% of its dry weight. Barring the unlikely event that the same ribosome actually serves in the synthesis of several distinct protein species, sizes significantly below the calculated values are possible only if a less bulky machinery makes proteins. Even a single ribosome surrounded by a membrane and a wall would occupy a sphere of 57 nm in diameter.

METABOLISM AND PHYSIOLOGY OF CONVENTIONAL BACTERIA

Dan G. Fraenkel
Department of Microbiology and Molecular Genetics
Harvard Medical School

The term "nanobacteria" reaches from asteroid fossil to agent of kidney stones, with the common thread of small size. To place the discussion in context, the following items survey the metabolism and physiology of bacteria as we know them, and, other than for item 7, size is not addressed. Bacteria are small, complex objects made of macromolecules, including proteins, which carry out chemical transformations and use simpler components to autonomously form more of themselves in a geometric manner. They have an inside, a membrane, and usually a wall, and their replication depends on coded information. Each term—size, complexity, autonomy, replication, wall, genome, metabolism, etc.—needs qualification, but that is not the present task, nor is discussion, other than implied, of how it all came about or where it is leading.

1. **Synthesis of the monomers.** *Primary carbon assimilation.* In nature new cells are ultimately made from inorganic materials, e.g., CO_2, NH_3, H_2S, etc. The ability to derive all carbon from CO_2 ("autotrophy") is widespread and perhaps a property of early cells. In that context, metabolism is a network of ca. 20 reactions connecting a few key compounds—sugar-phosphates, acetate, pyruvate, oxalacetate, and α-ketoglutarate (Figure 1), and primary carbon assimilation pathways are ways of contributing these compounds: in methanogens, a linear route comprising a handful of unique reactions, and more commonly, various cyclic routes such as the Calvin pathway.

The monomers. A complete chart of intermediary metabolism (see ref. 4) is daunting, but can be thought of as built up from Figure 1, with the key intermediates serving as starting materials for dedicated routes to the amino acids of the proteins, the constituents of polysaccharides, nucleic acids and lipids, and to cofactors. As seen in Figure 2, certain routes are short: glutamate is a single step from alpha-ketoglutarate, the reductive amination also introducing $-NH_2$; some are a little longer, serine by three steps and then introduction of –SH giving cysteine, and even the longest is only 10-15 reactions.

This adds up to ca. 120 reactions, including nucleotide interconversions. However, the various pathways share cofactors such as NAD^+, TPP, coenzyme-A, etc., which need to be made, too, with likely >150 additional reactions for the purpose.

2. **Energy.** Energy is needed (i) for provision of reductants for biosynthesis (if H_2 is not available) and of inorganic ions at appropriate reduction state (sulfate to H_2S, N_2 to NH_3, etc.), (ii) for kinetic activation of many steps, and (iii) to drive reactions whose equilibria are wrong (e.g., protein synthesis). Many of

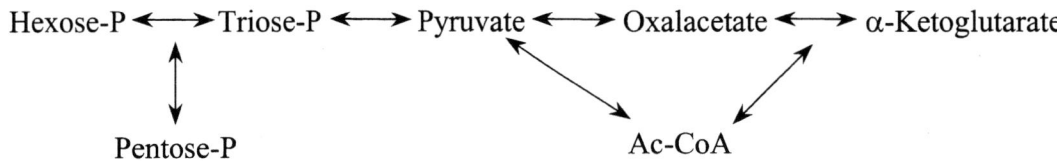

Figure 1. The basic framework of intermediary metabolism.

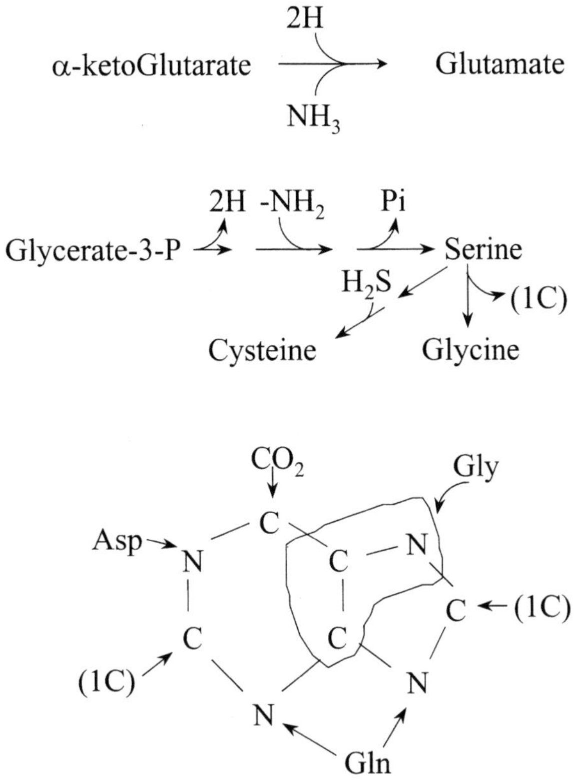

$$\alpha\text{-ketoGlutarate} \xrightarrow[\text{NH}_3]{2H} \text{Glutamate}$$

$$\text{Glycerate-3-P} \xrightarrow{2H} \xrightarrow{-NH_2} \xrightarrow{Pi} \text{Serine} \rightarrow (1C)$$
$$\text{H}_2\text{S} \swarrow \qquad \downarrow$$
$$\text{Cysteine} \qquad \text{Glycine}$$

Figure 2. Examples of monomer biosynthesis. The third item shows the source of atoms in purines.

the reactions directly use ATP or pyrophosphate but ultimately depend on ion gradients across the cell membrane formed by primary energy conservation devices. In turn, ion gradients energize an ATP synthase of ca. 10 subunits (Figure 3).

3. **Polymers.** For RNA, a polymerase (<10 genes). For DNA, a polymerase, plus a packaging and replication machinery (20 genes?). For protein, the tRNA's, amino acid activating enzymes, ribosomes with their associated factors (100 genes?).

4. **Membrane.** Membrane lipids are derivatives of glycerol (1 reaction from the central network) together with fatty acid esters or long chain ethers from acetate plus various head groups (ca. 20 genes?) Apart from primary energy conservation, the membrane also serves to (i) keep metabolites from dilution by the outside, (ii) exclude toxic materials, and (iii) concentrate materials from outside. The conservation function is partly met by the membrane being easily permeable only to very small uncharged molecules (H_2O, NH_3, and CO_2). Other materials require transport mechanisms, some being energy linked (Figure 4).

5. **Wall.** Life of normal sized bacteria in dilute solutions requires a wall to protect cells from lysis by osmotic pressure differences. Peptidoglycan (Figure 5), a cross-linked polymer based on glucosamine chains and amino acid bridges, often contributes this function; the rigid structure also requires a remod-

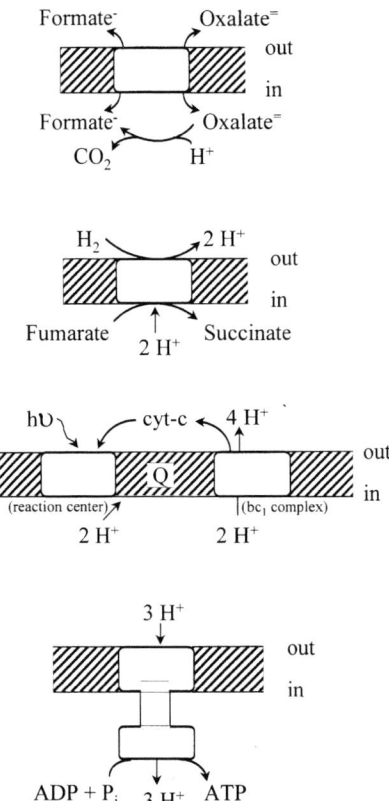

Figure 3. Examples of generation of protonomotive force (PMF) by (top) antiport; (next) electron transport; (next) photosynthesis; and (bottom) ATP synthesis by the ATPase.

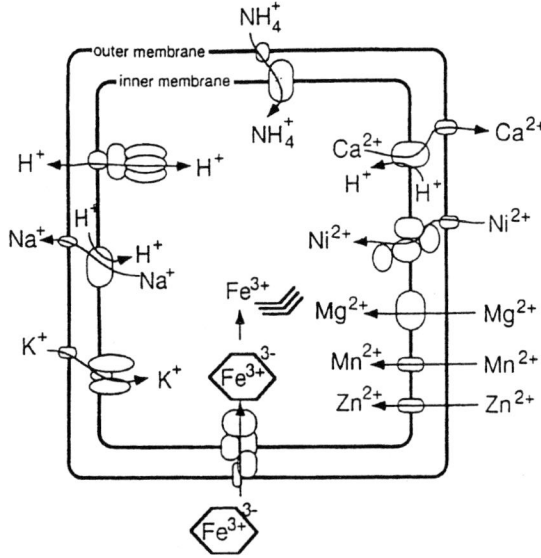

Figure 4. Cation transport in *E. coli*. Reprinted, by permission, from *Escherichia coli and Salmonella*, edited by F.C. Neidhardt et al. (1996), p. 1092. Copyright © 1996 by ASM Press.

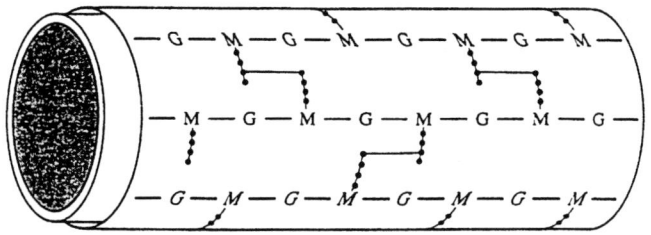

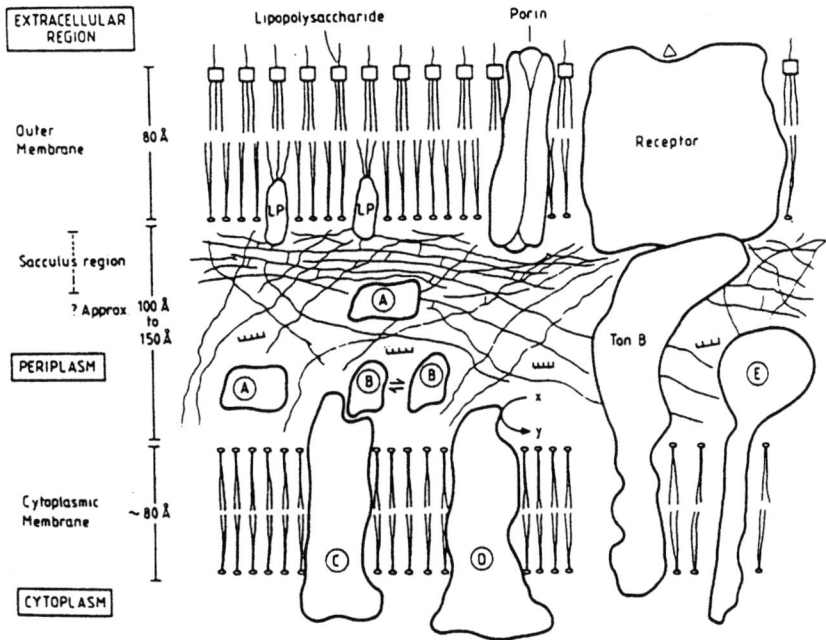

Figure 5. Peptidoglycan (top) and the gram-negative envelope (bottom). Reprinted, by permission, from *The Physiology and Biochemistry of Prokaryotes* by D. White (1995), pp. 14 and 18. Copyright © 1995 by Oxford University Press.

eling mechanism to avoid vulnerability during cell division (20 genes?). Bacterial envelopes are commonly of the complex gram-negative type (Figure 5), and, as a whole, diverse in their components.

6. *Heterotrophic metabolism*. *Catabolic pathways*. The best-known bacteria are not autotrophs but heterotrophs, needing an organic carbon source that is transformed by a catabolic pathway into the central network (Figure 6). Then, carbon intermediates and reductants come from the organic substrate. Heterotrophic bacteria of even moderate versatility, such as *E. coli*, may use dozens of different carbon and nitrogen sources, with hundreds of (nonessential) genes for this purpose.

Preformed monomers. The availability of exogenous monomers allows dispensability of biosynthetic pathways, too. Thus we cannot make and must eat the "essential" amino acids (those with longer

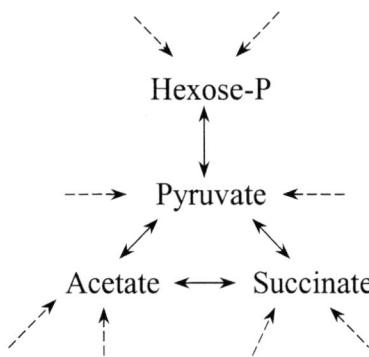

Figure 6. Catabolism.

pathways) and vitamins, and certain bacteria that live in tissues require even more. This might save ca. 100-200 genes, but transport systems would be needed.

Energy from catabolism. Certain catabolic routes, such as the widespread Embden-Meyerhof glycolytic pathway (Figure 7), which is used by many bacteria with access to carbohydrates, also contribute ATP by cytoplasmic reactions ("substrate level phosphorylation"); when growth depends solely on the latter reactions, the ATPase energizes the membrane rather than vice versa. Catabolism without respiration can involve massive wasting of organic metabolites as fermentation products. How-

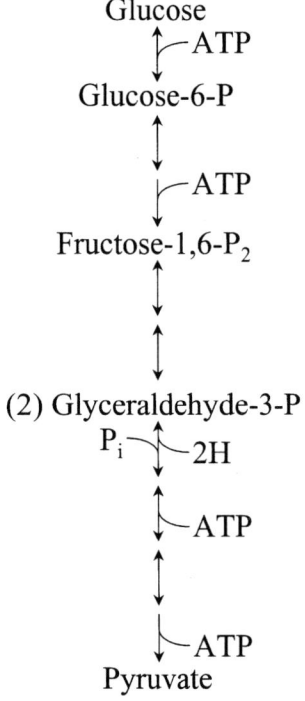

Figure 7. Embden-Meyerhof glycolytic pathway.

ever, in most heterotrophs, as in autotrophs, ATP derives from anaerobic or aerobic respiration, the latter pathways being similar to those of our mitochondria.

7. *How much is needed?* The essential set of genes for autonomous growth of modern bacteria in a minimal medium with a single organic carbon source or CO_2 is, from the above crude estimates, 400 or 500, and, as mentioned, fewer in an enriched medium. *E. coli* contains 4,300 genes (4×10^{-15} g DNA) and when growing in minimal medium with glucose has a volume of ca. 1 $(\mu M)^3$ and a dry mass of 250 $\times 10^{-15}$ g. An organism of 1/4 of those linear dimensions and hence ca. 2% of the mass (in line with sizes of "nanobacteria" cited by the organizers) would need to have less DNA than *E. coli*, and, with the minimal 400 gene complement, DNA would compose ca. 1/4 of its mass—not unreasonable. But an organism 1/10 of *E. coli* on the side and hence 1/1000 of the mass would already exhaust it with 200 average size genes. The size limitation is far stricter, because at least membrane is obligatory and, as pointed out by de Duve, ribosomes alone are a significant constraint.

Unlike genes, gene products need not be in equal amount: The synthesis of 1 *E. coli* cell from glucose may use three times its weight in glucose, or 4.2×10^{-9} mmoles; its 5% of dry weight as glycine in protein amounts to 1.7×10^{-10} mmoles and a cofactor at 10 μM in the soluble pool is ca. 3.5×10^{-15} μmoles. The net fluxes in the three enzymic pathways, 1 catabolic and 2 biosynthetic, therefore differ over a 10^6 range.

Furthermore, the amount of enzymes (their Vmax's) to provide adequate metabolite concentrations depends on growth rate, so if that value were 1/day instead of 1/h the cell might contain proportionately less enzyme. Both considerations—relative use of different pathways, and growth rate—suggest that small and slowly growing cells might have much less than the nominal 0.5 fraction of mass as catalytic protein.

This saving of protein has limitations. Although protein synthesis is related to need (fewer ribosomes in slower growth), the factor of gene expression is not controlled over a range of 10^6, many enzymes are in large excess, and enzymes vary greatly in their catalytic constants, with values usually higher for catabolic than anabolic reactions. Another constraint is seen from calculation of average number of molecules per cell: a nominal "nanobacterium" with 3×10^{-15}g protein, 4×10^{-15} ml cytoplasmic volume and 400 different polypeptides of size 40,000 could contain an average of 100 molecules and 30 μM of each one. Although the actual range of values could be large, values of less than one enzyme molecule per cell are possible only for steps whose products are in excess.

It should also be recalled that metabolic fluxes are not always tightly coupled to growth (rapid fermentations can proceed in non-growing cells), so that even "non-viable" bacteria might have significant complements of normally active pathways.

These several points suggest that even apart from physical constraints related to mechanisms of replication, the minimal size of conventional bacteria might be a few percent that of *E. coli* on glucose. For even smaller size, one could speculate about less-evolved cells with fewer constituents and simpler ribosomes, or about more-evolved ones, with multifunctional polypeptides of higher catalytic constants than presently common, hence little genome and protein but still of conventional design.

References

1. D. White, *The Physiology and Biochemistry of Prokaryotes*, Oxford University Press, New York, 1995, pp. 14, 18. (This is a useful introduction to bacterial metabolism and physiology from which figures were taken or adapted.)
2. F. Harold, *The Vital Force: A Study of Bioenergetics*, W.H. Freeman and Co., New York, 1986.
3. F.C. Neidhardt, et al., *Escherichia coli and Salmonella*, ASM Press, Washington, 1996, p. 1092.
4. Boehringer-Mannheim, *Biochemical Pathways,* G. Michal, ed., 3rd Ed., 1992. (A chart that shows it all at a glance.)

A BIOPHYSICAL CHEMIST'S THOUGHTS ON CELL SIZE

Peter B. Moore
Departments of Chemistry and Molecular Biophysics and Biochemistry
Yale University

Abstract

Back-of-the-envelope calculations having to do with genetic complexity and intracellular DNA concentration suggest that actively growing cells significantly smaller than those of *Mycoplasma genitalium* (diameter ~0.3 µm) are unlikely to exist on Earth today. In addition, the limitations on genome size that accompany small cell size guarantee that cells of that size will have to depend on larger, more complex organisms for supplies of most of the small organic molecules required for their growth. Hence, a biosphere consisting entirely of ultra-small organisms is highly implausible.

Introduction

Cells differ enormously in size. Some of the neurons in the human brain have processes meters in length, and microbiologists routinely study free-living, single-celled organisms that have linear dimensions measured in micrometers. Viruses can be almost two orders of magnitude smaller than that. In light of the recent report of McKay et al. (1996) suggesting that objects found in a meteorite of martian origin, which are the size of a large virus (diameter ~0.1 µm), might derive from ancient organisms, it is interesting to ask how small an object can be and still be a cell. For these purposes, I take a cell to be a membrane-enclosed entity that (1) replicates itself autonomously, and (2) is actively growing.

My perspective on this problem is that of a biophysical chemist, not that of a systematic microbiologist. Thus, while I am in no position to testify about the sizes and "lifestyles" of very small terrestrial organisms, I can comment on some of the physical issues that small cells confront.

Critical Assessment of the Issues

As it happens, I have spent much of my career working with *Escherichia coli*, a bacterium whose linear dimensions are of the order of micrometers, and it is useful to start by reviewing its capabilities and chemical composition. *E. coli* lives primarily in the intestinal tracts of mammals, where the temperature is a comfortable 37° C, and both water and its hosts provide nutrients. Even though this environment is nutritionally rich, *E. coli* has remarkable metabolic capabilities. It thrives in media consisting of mineral salts and glucose, a diet far too sparse to support human life, for example. *E. coli* will grow on similar media in which glucose is replaced by many other small organic molecules, including acetate, which is about as simple a substrate as it can manage. Only photosynthetic organisms, which will grow on salts, CO_2, and light, can make do with significantly less.

How big is such an *E. coli* cell, and what does it contain? Answers to both questions may be found in Watson's classic text, *Molecular Biology of the Gene* (Watson, 1965). Table 1 is a version of a table in Watson's book that I revised using information gleaned from the sequence of the *E. coli* genome (Blattner et al., 1997). The cell whose contents are outlined in Table 1 is a cylinder 2 µm long and 1 µm in diameter. Its volume is thus about 1.6 µm^3 and its mass is about 1.7×10^{-12} g. The composition given

Table 1 The Composition of a Typical *E. coli* Cell

Substance	% of mass	MW	# of copies	# of kinds
Water	70	18	4×10^{10}	1
Inorganic ions and small molecules of all other kinds	7	~145	5×10^8	750
DNA	1	3×10^9	4	1
RNA	6			
16S rRNA		0.5×10^6	30,000	7
23S rRNA		1.0×10^6	30,000	7
5S rRNA		4.0×10^4	30,000	7
tRNA		2.5×10^4	300,000	86
mRNA		~1.0×10^6	~1,000	~1,000
Proteins	15	~30,000	5×10^6	4,288

SOURCE: Adapted from Watson (1965); data from Blattner et al. (1997).

is that expected for rapidly growing cells, and that explains why it contains several copies of its genome, not just one.

It is important to realize that the number of ribosomes in bacteria like *E. coli* is tightly regulated so that if all the ribosomes present in a cell constantly make protein at the maximum possible rate (10-20 amino acids incorporated per ribosome per second), the protein mass in the cell will have just doubled by the time cell division begins. The number of copies of each of the enzymes involved in intermediary metabolism is similarly regulated to balance supply and demand. Hence the macromolecular composition of these cells depends on their overall growth rate, which in turn is coupled to the mix of small organic molecules available in the environment. Finally, it should be noted that the water content of these cells, 70% by weight, is typical of actively metabolizing cytoplasm.

A cell a tenth the volume of *E. coli* is easy to envision. It might have 2 copies of a genome that is about the same size, which would imply an interior DNA concentration 5 times that of *E. coli*. This means that there would not quite be room in such a cell for the 3,000 ribosomes and 5×10^5 protein molecules it would require, and still maintain a water content of 70%. However, if the organism's metabolism were adjusted so that its generation time was a bit longer than that of *E. coli*, all would be well. There is a limit to how far this kind of balancing of generation times and macromolecular compositions will take you. The reason is that while the number of copies of each kind of macromolecule a cell requires scales approximately with its volume, the amount of genome it takes to encode them does not. The consequences become obvious when we consider cells whose volumes are 1% that of *E. coli*. A cell that size has a mass of only ~10^{10} daltons. If it must do all the things *E. coli* can, it needs a genome the same size, i.e., one that weighs 3×10^9 daltons. If it also has to be 70% by water weight, there is no room for anything else.

It is instructive to examine Table 2, which compares the DNA concentrations in three DNA-containing objects, the interior volumes of which stand approximately in the ratio of 10,000:100:1: *E. coli*, *Mycoplasma genitalium*, a small mammalian parasite, and the head structure of bacteriophage T4, which is a virus. Even though the genome of *M. genitalium* is significantly smaller than that of *E. coli*, as one anticipates it must be, it isn't two orders of magnitude smaller, and consequently the DNA concentration in *M. genitalium* is much higher than in *E. coli*. The DNA concentration in the T4 head, which is higher still, approaches that of pure, hydrated DNA.

It is important to realize that the DNA in T4 heads is metabolically inert, and is intended to be so.

Table 2 DNA Concentration as a Function of Organism Size

Organism	Radius (μm)	Volume (μm³)	Genome MW	# of copies	gDNA/ml
E. coli	0.73	1.6	3×10^9	4	0.013
M. genitalium	0.15	1.3×10^{-2}	4×10^8	~2	0.100
Phage T4	0.04	2.4×10^{-4}	1×10^8	1	0.690

NOTE: The data for *E. coli* are taken from Watson (1965). The information for Mycoplasma comes from Morowitz (1992), and the bacteriophage T4 data come from Stryer (1981). For all three, it is the interior volume that is of interest. The interior volume of *M. genitalium* was computed allowing 0.005 μm for the thickness of its surrounding lipid bilayer. No correction for membrane thickness was made for *E. coli*. The volume Stryer provides for the head of bacteriophage T4 is that of its interior cavity.

Viruses don't do anything; they are simply genome transport modules. Not only that, it is impossible for the DNA in T4 heads to be replicated or expressed in situ because there isn't room for the macromolecules that catalyze these processes. T4 DNA becomes metabolically active only when injected into bacterial cells where the total DNA concentration is much lower. I take it as axiomatic that DNA packed as densely as it is in bacteriophages will always be inert metabolically. If this is so, unless cellular genomes exist that are even smaller than that of *M. genitalium*, cells significantly smaller than *M. genitalium* cells are impossible. On the basis of what has been learned from the sequence of the genome of *M. genitalium*, this looks unlikely.

As Table 3 indicates, the genome of *M. genitalium* encodes only a tenth as many proteins as that of *E. coli*, and it is the absence of these "missing" proteins from its cytoplasm that allows *M. genitalium* cells to devote as large a fraction of their interior volumes to DNA as they do. However, this reduction in protein inventory has a price. Compared to *E. coli*, *M. genitalium* is a physiological cripple. For example, it lacks almost all the proteins required of intermediary metabolism, and hence must find virtually all the small organic molecules it needs in its environment. It is hard to imagine an organism with such limited capabilities surviving, except as a parasite.

Table 3 The Protein Inventories of *E. coli* and *M. genitalium*

Functional Class	Number of Genes	
	E. coli	M. genitalium
Regulation	178	7
Structure	237	17
Phages and other inserted elements	87	0
Transport and binding	427	34
Energy metabolism	243	31
DNA replication, etc.	115	32
Transcription	55	12
Translation	191	101
Intermediary metabolism	658	37
Other cell processes	188	21
Other enzymes and identifiable genes	277	27
Unknown	1,632	152
Total	4,288	471

SOURCE: These data are taken from Blattner et al. (1997) and Fraser et al. (1995).

Marine ultramicrobacteria, which are also unusually small, have "design specifications" consistent with the arguments made above. *Sphinogomonas sp.*, which appears to be typical, has a cell volume 4 times that of *M. genitalium*, and a genome a quarter the size of *E. coli's* (Schut et al., 1997). A genome that size is big enough to encode enough enzymes to explain the intermediary metabolism of these organisms, and intracellular DNA concentration is certainly not a problem. The free-living nanobacteria that Kajander and his colleagues have isolated from mammalian cell cultures are harder to understand at this point (Kajander et al., 1997). The volumes of some of the objects seen in such cultures are a tenth that of *M. genitalium* (!), and it is hard to see how anything could grow autonomously with a genome much smaller than that of *M. genitalium*. The biochemistry and genomics of these organisms are bound to be very interesting, but one anticipates that when the dust clears, it will be found that they too "fit inside the envelope."

In this connection, it is worthwhile asking whether one can imagine organisms with genomes even smaller than that of *M. genitalium*? Perhaps. Experiments done in the 1970s suggest that modern proteins might be larger than their functions absolutely require. For the sake of argument, assume that all the genes in our putative nanocell genome are half the size of those found in *M. genitalium* or *E. coli*, which, it should be pointed out, are about the same size (~ 30 kD, on average). Let us also postulate that this genome is single-stranded DNA, not double-stranded, and that it replicates by a process in which only a partial copy of its Watson-Crick complement ever exists. For good measure, let us also arbitrarily dispense with 100 of the genes found in the *M. genitalium* genome. The molecular weight of the DNA genome that emerges is about 0.7×10^7, and if we allow the intracellular concentration of DNA to rise to 0.2g/ml, which is twice that found in *M. genitalium*, the estimate for interior cell volume that emerges is $\sim 10^{-3}$ μm^3, which is a factor of 10 less than that of *M. genitalium*. Omitting an allowance for the surrounding bilayer, the radius of the interior of such a nanocell would be 0.06 μm, which is close to the size of the cells Kajander and his colleagues have identified and of the objects that McKay et al. (1996) have suggested might be traces of martian cells.

Several comments are in order. First, in order to arrive at a "design" for this 0.06 μm cell, we have invoked a biochemistry unlike any known in modern cellular life. A cell that is even smaller would require even more radical departures, which I lack the imagination to envision. Second, whatever else this nanocell can do, it is not going to grow on a medium consisting of mineral salts and glucose. Something else, which is far larger and more complicated, will have to do the biochemical "heavy lifting" that makes the survival of such organisms possible. To put it another way, a biosphere consisting entirely of organisms of the size and metabolic complexity of *E. coli* is easy to envision; it may once have existed on this planet (Schopf, 1983). A biosphere consisting of organisms the size of *M. genitalium* or smaller looks impossible. Third, these considerations suggest that it is appropriate to scrutinize all reports of cells significantly smaller than *M. genitalium* with great care. Are these entities really cells, or are they spores, or merely components of cells? Can they grow without a major reorganization of their morphologies and internal composition? Fourth and finally, wise biologists never say "never." Maybe cells significantly smaller than those we have been considering really do exist. If they do, the one thing we can be sure is that their biochemistry will be a lot different from anything we know today.

References

1. Blattner, F.R., G. Plunkett, C.A. Bloch, N.T. Perna, V. Burland, M. Riley, J. Collado-Vides, J.D. Glasner, C.K. Rode, G.F. Mayhew, J. Gregor, N.W. Davis, H.A. Kirkpatrick, M.A. Goeden, D.J. Rose, B. Mau, and Y. Shao. 1997. The complete genome sequence of *Escherichia coli* K-12. *Science* **277**: 1453-1474.

2. Fraser, C.M., J.D. Gocayne, O. White, M.D. Adams, R.A. Clayton, R.D. Fleischmann, C.J. Bult, A.R. Kerlavage, G. Sutton, J.M. Kelley, J.L. Fritchman, J.F. Weidman, K.V. Small, M. Sandusky, J. Fuhrmann, D. Nguyen, T.R. Utterback, D.M. Saudek, C.A. Phillips, J.M. Merrick, J.-F. Tomb, B.A. Dougherty, K.F. Bott, P.-C. Hu, T.S. Lucier, S.N. Peterson, H.O. Smith, C.A.I. Hutchinson, and J.C. Venter. 1995. The minimal gene complement of *Mycoplasma genitalium*. *Science* **270**: 397-403.
3. Kajander, E.O., I. Kuronen, K. Åkerman, A. Pelttari, and N. Çiftçioglu. 1997. Nanobacteria from blood, the smallest culturable autonomously replicating agent on Earth. *Proceedings of SPIE* **3111**: 420-428.
4. McKay, D.S., E.K. Gibson, Jr., K.L. Thomas-Keprta, H. Vali, C.S. Romanek, S.J. Clemett, X.D.F. Chillier, C.R. Maechling, and R.N. Zare. 1996. Search for past life on Mars: possible relic biogenic activity in Martian meteorite ALH84001. *Science* **273**: 924-930.
5. Morowitz, H.J. 1992. *Beginnings of Cellular Life,* New Haven: Yale University Press.
6. Schopf, J.W., ed. 1983. *The Earth's Earliest Biosphere,* Princeton, N.J.: Princeton University Press.
7. Schut, F., J.C. Gottschal, and R.A. Prins. 1997. Isolation and characterization of the marine ultramicrobacterium Sphingomonas sp. strain RB2256. *FEMS Microbiol. Rev.* **20**: 363-369.
8. Stryer, L. 1981. *Biochemistry,* 2nd Ed., San Francisco: W.H. Freeman & Co.
9. Watson, J.D. 1965. *Molecular Biology of the Gene,* 1st Ed., New York: W.A. Benjamin.

CORRELATES OF SMALLEST SIZES FOR MICROORGANISMS

Monica Riley
Woods Hole Marine Biological Laboratory

Abstract

The size of Earth-bound bacteria is dictated by a number of factors, the most important of which is the growth rate. If we postulate that ancient cells had very slow growth rates and therefore required few ribosomes, and if they lived in nutritionally rich surroundings, thus requiring few biosynthetic enzymes, they could conceivably be as small as a 200-nanometer-diameter sphere. More drastic scenarios, such as supposing a one-polymer information system or a sharing system for part-time genes, could reduce the size requirement further.

Introduction

We are asking ourselves what, from the point of view of traditional microbiology and biochemistry, are the factors that dictate a lower limit to the size of microorganisms of the kind that are present in our life system? The question has different answers depending on the composition of the environment of the microorganism, but it is possible to consider in turn major factors that influence the estimate.

The Size and Contents of an Average Gram-Negative Organism

Escherichia coli is a typical gram-negative rod bacterium. Its dimensions are those of a cylinder 1.0-2.0 micrometers long, with radius about 0.5 micrometers. Another gram-negative rod, less metabolically independent than *E. coli*, is *Hemophilus influenzae*, which has half the length and diameter. Bacteria such as mycoplasma, which have a more modest metabolic capability, are even smaller.

The Effect of Growth Conditions on Cell Constituents

If the environment of the culture of microorganisms is rich, the growth rate will be fast relative to the growth rate in minimal medium. The distribution of contents of *E. coli* changes with growth rate (1-4) as shown in the Table 1.

The cytoplasm, which contains the soluble proteins and the protein synthesis apparatus, is the dominant space-filling aspect of the microbial cell. Ribosomes dominate the cytoplasm and are major space-occupying cellular elements. Compared to the cytoplasmic proteins and the ribosomes, cellular

Table 1 Cell Contents at Fast and Slow Growth Rates (Percent by Weight)

Macromolecule	Fast Growth (<30 min doubling time)	Slow Growth (>150 min doubling time)
DNA	3%	4%
RNA	35	14
Protein	60	80
Cell size	Larger	Smaller

ingredients that do not occupy much volume are the DNA, the cell wall, and the membranes with their component transport systems. Nucleoids are composed of compacted DNA neutralized with either positive ions or basic proteins. The DNA is so compacted that within the nucleoid the concentration of DNA reaches 10-20% and constitutes only about 10% of cell volume.

Distribution of Genetic Determinants of *E. coli* K-12

The genetic determinants of *E. coli* K-12 distribute as follows:

Enzymes	1,300
Transporters	500
Regulators	500
RNAs	115
Structure	200
Factors	25
Unknown	1,800

The numbers and kinds of genes expressed at any one time are determined by the composition of the environment. Enzymes are repressed unless needed, induced when needed. Numbers of ribosomes are determined by environmental conditions (1,3).

Essentials of Conceptually Pared Down Average Cell Like *E. Coli*

Not So Much DNA

A minimal free-living cell need not have as much DNA as *E. coli* does because *E. coli* is able to live in many different circumstances and carries genes for many more than the minimum number of enzymes required at any one time. If one is thinking about organisms living in a biochemically rich soup or organisms like parasitic bacteria that live inside cells and draw on the processes and contents of inhabited cells, the organism could be smaller than if it were a free-living cell that makes all its own substance from scratch. Presently known smallest bacterial genomes are around 600 Kb of DNA.

Ribosomes

A major element determining the size of bacteria is their ribosomes. Under different growing conditions, *E. coli* has 10,000-60,000 ribosomes. Each ribosome contains two RNA molecules and a very large number, 52, of different proteins: 21 in the 30S subunit, 31 in the 50S subunit. The protein complex called L7/12 is represented 4 times in each ribosome, but probably the rest of the ribosomal proteins are present only once per ribosome (5). Other proteins are involved in directing the correct assembly of the components of the ribosomes.

Thus, ribosomes are a prominent feature determining minimum cell size in any life system using ribosomes as an obligatory element. However, we can imagine circumstances not requiring as many ribosomes as *E. coli* has. The number of ribosomes is steeply dependent on growth rate; namely, the slower the growth rate, the fewer ribosomes needed (1,2,4). If we postulate that growth rates of very early, very small cells could have been very slow, ribosomal content could have been correspondingly lower than today's organisms. If we postulate extremely slow growth, can we postulate as few as two ribosomes per cell? This question reaches outside our known world and cannot be answered knowledgeably today.

Proteins

Although some proteins of *E. coli* are the ribosomal proteins, most are soluble enzymes. Other proteins are integral membrane proteins allowing for cellular localization and carrying out transport functions. Membrane and wall components and also cellular appendages such as flagellae and pili account for a very small part of the weight of a bacterium, so we will leave them aside. Soluble cytoplasmic proteins are chiefly enzymes of metabolism and regulators of all cell processes. The numbers of necessary metabolic enzymes and their regulators required for life depends entirely on the genetic capacities and the metabolic capabilities of the cell.

Metabolism

Quasi-redundancy. *E. coli* is profligate in its metabolic enzymes. It squanders its information content on multiple enzymes. There are over eighty examples of small-molecule metabolic reactions that are carried out by more than one separately encoded isozyme (6). In addition, there are multiple large-molecule enzymes such as polymerases, helicases, repair enzymes. Whereas *E. coli* is a versatile organism able to draw on isozymes of slightly different characteristics to guarantee life under different circumstances, our theoretical very small, very early organism may not have been versatile. A minimal organism might have only one enzyme per reaction, and the number of reactions catalyzed may have been small.

"Totipotency." *E. coli* possesses flexibility and metabolic alternatives as a consequence of containing the enzymes for many metabolic capabilities, only some of which are working under any environmental condition or at any point in time. A minimal organism could make do with minimal metabolic capacity.

Degradation and Energy. All major degradative pathways are in *E. coli*: glycolysis, pentose shunt, Entner-Douderoff pathway, TCA cycle, and glyoxalate shunt. *E. coli* has the ability to live anaerobically as well as aerobically, with many other electron acceptors besides oxygen (among them nitrate, sulfate). *E. coli* can derive energy and make ATP from many oxidizable substances by using either respiration with electron carriers, both aerobic and anaerobic, or fermentation involving only reduced and oxidized organic compounds. *E. coli* contains the complex formate dehydrogenase system and hydrogenase, enabling it to derive energy from such simple molecules as hydrogen or formate, the simplest organic acid. *E. coli* can use as carbon and energy sources a wide variety of substrates (although it does not fix CO_2). Thus, the degradative and energy metabolism of *E. coli* is much more complex than it would be for a minimal organism.

Simple ways a minimal organism might use to gain energy are by fermentation, using one organic molecule as electron donor and another molecule as acceptor (as a chemoorganotroph) or by oxidizing inorganic substances such as H_2S (as a chemoautotroph).

Anabolism. The synthetic capacities of *E. coli* are also extensive. It can make all of its small-molecule building blocks such as amino acids, purines, pyrimidines, and cofactors, and for some pathways there are alternative routes using other enzymes. If the necessary building blocks for life were in the environment of a minimal cell, the synthetic metabolic capacity would not be needed. However, one would need transport mechanisms to bring the building blocks in and not leak out essential metabolites.

If, on the other hand, primitive environments did not supply preformed metabolites, then the machinery for CO_2 fixation, perhaps also for N_2 fixation and full anabolic capability would be needed. The complexity and need for more proteins could then rise back up to *E. coli* levels.

Counting Up

E. coli contains over 1,000 known soluble enzymes, and a high proportion of the 1,800 unknown ORFs [open reading frames (genes likely to code for proteins)] are expected to be enzymes. In a minimal cell with minimal small-molecule metabolism and minimal macromolecule mechanisms, maybe the enzyme count could be reduced to as little as 300.

Reducing the number of proteins also reduces the number of ribosomes needed to synthesize the proteins. Primitive cells with slow growth characteristics could have had far fewer ribosomal RNA molecules and could have used many fewer individual proteins to build primitive ribosomes. Because with slow growth conditions protein (the sum of enzymes and ribosomal protein) is the major component of a cell, reduction in number of unique proteins seems to be a major factor in determining the complexity and size of the cell.

Macromolecule processes such as protein synthesis and DNA replication could have been even simpler in a very small and very ancient cell. What are the minimal needs for these processes in an "RNA world"(7)? Can the machinery be reduced to a much simpler version, eliminating altogether DNA and its replication? Do RNA and protein constitute a two-polymer system? Can we imagine a one-polymer system in which RNA is possessed of adequate catalytic powers to be able to substitute for proteins, or where proteins can replicate themselves without a polynucleotide template?

Envelope

Another factor to think about is the increased surface volume of small cells as compared to large. Although one expects the structures to be much simpler than the envelope of gram-negative bacteria, still a minimal cell would have a higher proportion of its substance in its membrane and wall fractions.

Summary

Can we think, then, about a very early cell with an enclosing membrane that has fewer transporters than at present, only partially controls transport in and out, and contains only a few ribosomes? It would contain only about 300 proteins including enzymes, a small number of generalized regulators, and perhaps no DNA, only RNA performing tasks of coding, processing, and replication. Such a cell would need to have a volume of at least 200 nanometers^3. To imagine smaller cells than this requires a higher level of speculation, such as having only part-time genes that come and go, or having a one-polymer system where RNA is possessed of adequate catalytic powers to be able to substitute for proteins (7), or where proteins can replicate themselves without a polynucleotide template.

References

1. Ingraham, J.L., O. Maaloe, and F.C. Neidhardt. 1983. *Growth of the Bacterial Cell.* Sunderland, Mass.: Sinauer Associates, Inc.

2. Bremer, H., and P. Dennis. 1996. Modulation of chemical composition and other parameters of the cell by growth rate. Pp. 1553-1569 in *Escherichia coli and Salmonella,* 2nd Ed., F. Neidhardt, R. Curtiss III, E.C.C. Lin, J. Ingraham, K.B. Low, B. Magasanik, W. Reznikoff, M. Riley, M. Schaechter, and H.E. Umbarger (eds.), Washington, D.C.: ASM Press.
3. Maaloe, O., and N.O. Kjeldgaard. 1966. *Control of Macromolecular Synthesis.* New York: W.A. Benjamin Co.
4. Ingraham, J., and A.G. Marr. 1996. Effect of temperature, pressure, pH and osmotic stress on growth. Pp. 1570-1578 in *Escherichia coli and Salmonella,* 2nd Ed., F. Neidhardt, R. Curtiss III, E.C.C. Lin, J. Ingraham, K.B. Low, B. Magasanik, W. Reznikoff, M. Riley, M. Schaechter, and H.E. Umbarger (eds.), Washington, D.C.: ASM Press.
5. Noller, H.F., and M. Nomura. 1996. Ribosomes. Pp. 167-186 in *Escherichia coli and Salmonella,* 2nd Ed., F. Neidhardt, R. Curtiss III, E.C.C. Lin, J. Ingraham, K.B. Low, B. Magasanik, W. Reznikoff, M. Riley, M. Schaechter, and H.E. Umbarger (eds.), Washington, D.C.: ASM Press.
6. M. Riley, and B. Labedan. 1996. *E. coli* gene products: Physiological functions and common ancestries. Pp. 2118-2202 in *Escherichia coli and Salmonella,* 2nd Ed., F. Neidhardt, R. Curtiss III, E.C.C. Lin, J. Ingraham, K.B. Low, B. Magasanik, W. Reznikoff, M. Riley, M. Schaechter, and H.E. Umbarger (eds.), Washington, D.C.: ASM Press.
7. Gesteland, R.F., T.R. Cech, and J.F. Atkins. 1998. *The RNA World,* 2nd Ed. Cold Spring Harbor, N.Y.: Cold Spring Harbor Laboratory Press.

MECHANICAL CHARACTERISTICS OF VERY SMALL CELLS

David Boal
Department of Physics
Simon Fraser University

Abstract

The smallest terrestrial cells have few structural elements: a fluid membrane to isolate the cell's contents, a very large molecule to carry its genetic information, and, often, a cell wall to offset the osmotic pressure of the cell's interior. There are few theoretical obstacles to constructing cells with radii as small as 50 nm using the same molecular materials as are found in 300 nm mycoplasmas. The energy required to bend a flat fluid membrane into the shape of a cell is comparatively small, such that closed spherical shapes are energetically favored for radii greater than about 20 nm, depending upon composition. Further, the membrane of a small cell could withstand the osmotic pressures typical of many bacteria without the aid of a cell wall. However, it would be difficult to pack a genetic blueprint with a hundred genes into a small cell using double-stranded DNA, whose rigidity permits only gentle curvature on 50 nm length scales; rather, a small cell would employ most flexible molecules such as RNA or single-stranded DNA.

Introduction

The human body contains about 10^{13} cells—perhaps a hundred times the number of stars in the Milky Way—although only about 200 different cell types are represented in this collection. A minimal set of mechanical components is present in each cell to perform such tasks as isolating its contents, maintaining its shape or, in some cases, facilitating its movement. The *chemical* composition of these *structural* components does not vary strongly from one cell type to another, permitting us to understand, in a somewhat systematic fashion, the architecture that nature has chosen for the cell. Small cells, such as bacteria, have a particularly simple construction:

- a fluid membrane (and possibly a cell wall) forming the cell boundary,
- an interior fluid region likely at higher pressure than the cell's immediate environment,
- at least one large molecule carrying the cell's genetic information.

Some questions that we might ask about the mechanical characteristics of these components are illustrated in Figure 1.

The properties of many of the cell's structural elements are known as a function of their size. For example, the filaments of the cytoskeleton (the molecular scaffolding that helps a cell organize its internal compartments and maintain its shape) display a resistance against bending that grows rapidly with their radius, just as rope is stiffer than string. Thus, we can predict, if crudely, the size of the cytoskeletal components needed for a cell to function under various conditions. In addition, limits or bounds exist on the minimum mechanical strength required of these components: for example, the fluid membrane enclosing the cell must possess a certain minimal resistance against rupture on a time scale appropriate to the cell's lifetime.

Because a given structural element may play several different roles in a cell, a limit based solely on

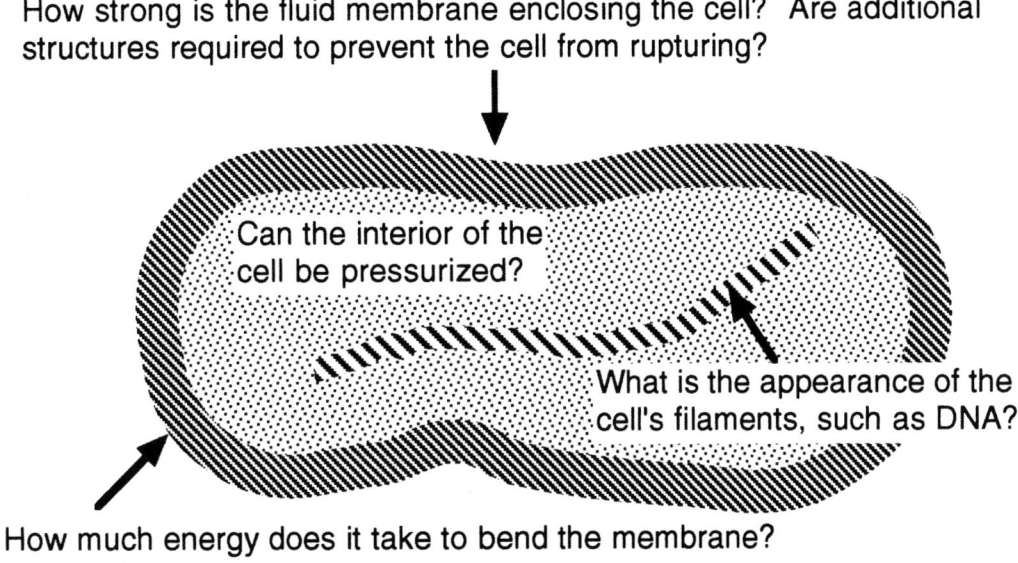

Figure 1. Some civil engineering issues facing the designer of very small cells.

a single mechanical characteristic may not truly reflect the complete architectural specifications of that element. As an illustration, a membrane with a molecular composition providing good rupture resistance may be so viscous that proteins are unable to diffuse readily within it. Thus, the actual molecular composition of the membrane reflects not only the strict limit on its mechanical strength, but also a softer constraint arising from the functionality of its constituents. Further, the limits are not inviolable and should be regarded more as challenges to Nature. The constraints that we obtain here assume only the most rudimentary architecture and the simplest chemical compositions. We make these assumptions in the belief that the smallest cells arise early in the history of a planet and have not had sufficient time to develop a complex architecture. However, there is nothing to prevent Nature from finding ingenious strategies to circumvent mechanical constraints that strictly apply only to the most structurally simple cells.

Viability of Very Small Cells

In this paper, we focus on just a few of the cell's mechanical properties: the resistance of the boundary membrane to bending and rupture and the elasticity of a cell's filaments. We then discuss the implications of these characteristics to the mechanical functionality of cells much smaller in size than typical terrestrial cells. Our benchmark is a structurally simple cell of radius 50 nm. We demonstrate that:

• In simple models of fluid membranes, the bending energy of a spherical shell is independent of its radius, so that it takes the same amount of energy to bend a flat membrane into a small spherical shell as a large shell. Whether such a cell is stable depends upon its energy compared to other configurations such as a flat disk with a free boundary. The creation of a hole or a free edge in a membrane requires an input of energy that is proportional to the length of the edge boundary. Except for very small membrane segments, it is energetically more favorable for a membrane with a free boundary to close up into a spherical shape, eliminating the boundary. The estimated minimum sphere radius arising from this argument is about 20 nm.

- There is a minimum stress that a membrane can tolerate before it ruptures on conventional time scales. Because the (surface) stress on a spherical shell is proportional to its radius, a small cell can tolerate higher internal pressures than can a large cell for a given membrane composition. Thus, a very small cell would not require a cell wall in order to function at the osmotic pressures typical of many bacteria.

- The bending resistance of a filament rises rapidly with its radius, so that thick filaments are relatively inflexible. Although a very small cell does not have sufficient volume to accommodate a conventional cytoskeleton (whose elements may be 10-25 nm in diameter), even a filament of double-stranded DNA would appear somewhat stiff on the scale of 50 nm. In order to code sufficient genetic information in a linear sequence, small cells would need very flexible molecules with perhaps half the mass per unit length of DNA, a requirement that is consistent with the idea that RNA or some other single-stranded molecule is the evolutionary precursor of DNA as the genetic template.

Detailed Analysis

Membrane Curvature

All cells are bounded by a plasma membrane consisting of a bilayer of dual-chain lipid molecules within which are embedded proteins and other molecules such as cholesterol. Bilayers are self-assembled structures whose equilibrium configuration is spatially flat if the molecular composition is the same within both layers. Such symmetric bilayers resist bending with an energy cost per unit area ε whose simplest parameterization is

$$\varepsilon = (\kappa/2)(1/R_1 + 1/R_2)^2 + \kappa_G/(R_1 R_2), \tag{1}$$

where the constants κ (bending rigidity) and κ_G (Gaussian curvature modulus) have units of energy [for a review of more complete descriptions of bilayer bending, building on the original approach of Helfrich (1973), see Lipowsky (1991)]. The quantities R_1 and R_2 are the two principal radii of curvature displayed in Figure 2. As an illustration, a sphere of radius R has $R_1 = R_2 = R$, while a cylinder has an infinite radius of curvature along the axis of cylindrical symmetry. To find the bending energy of a particular surface, one simply integrates ε over the entire surface: for example, a spherical shell has a bending energy of $8\pi\kappa + 4\pi\kappa_G$, independent of the shell's radius.

What is the magnitude of the bending energy for typical cells? Lipid bilayers in terrestrial cells are found to have $\kappa = 10\text{-}25\ k_B T$, where k_B is Boltzmann's constant and T is the temperature [see Evans and Rawicz (1990) and references therein]. The value of κ_G is much less well known, but is expected to have a similar magnitude as κ. With $\kappa = \kappa_G$, the energy of a spherical shell is $12\pi\kappa$. Considering only the contribution from κ, the bending energy of a spherical cell would be 250-600 $k_B T$. Although this is not really a large amount of energy (recall that $k_B T$ is roughly the kinetic energy of an atom in a gas), why would nature expend this energy to form a closed surface from an open bilayer sheet? To answer this question, we examine how a bilayer might rupture.

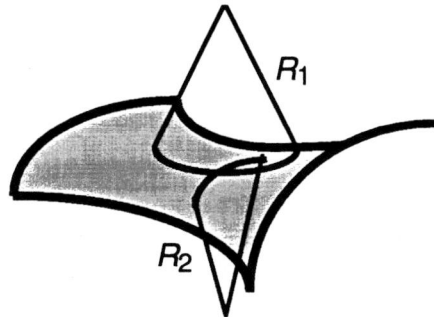

Figure 2. Principal radii of curvature for a saddle-like surface.

Membrane Rupture

The fluid membrane not only resists bending, but also resists in-plane stretching. Under tensile stress, the membrane first stretches then ruptures once the area has expanded a few percent beyond its unstressed value. The creation of a hole in a membrane likely involves reconfiguring the lipid molecules around the boundary of the hole in order to reduce contact between the aqueous medium surrounding the bilayer and the water-avoiding hydrocarbon chains of the lipid molecules, which are normally buried within the bilayer. In general, the orientation of the lipids at the hole boundary is energetically unfavorable compared to that of an intact bilayer, so that there is an energy penalty if the membrane has a hole or a free edge.

The boundary of the hole can be characterized by an edge tension λ (energy per unit length along the boundary), which has been measured to be in the 10^{-11} J/m range (for example, Fromherz, 1983); the measured values are larger than the minimum edge tension for membrane stability estimated from computer simulations of membrane rupture (Boal and Rao, 1992). For example, the edge energy of a flat disk of radius R_{disk} and perimeter $2\pi R_{disk}$ is $E_{disk} = 2\pi R_{disk}\lambda$. A membrane having this shape will be energetically favored over the closed sphere considered above ($E_{sphere} = 12\pi\kappa$ for $\kappa = \kappa_G$) if $R_{disk} < 6\kappa/\lambda$. If the disk and the sphere have the same surface area then $R_{sphere} = R_{disk}/2$ (see Figure 3). Thus we expect $R_{sphere} > 3\kappa/\lambda$ (after Fromherz, 1983). Using typical values of $\kappa \sim 15\ k_B T$ and $\lambda = 10^{-11}$ J/m leads to $R_{sphere} > 20$ nm, a bound whose exact value depends upon the membrane composition. Experimentally, one finds that pure bilayer vesicles (simple artificial cells in some sense) can be produced in the lab with radii as small as 30 nm (Fromherz, 1983; Frisken, 1998, private communication). Once the membrane has adopted a closed shape, the configuration could be further stabilized by the addition of lipids to the outer layer, thus reducing the strain in the bilayer.

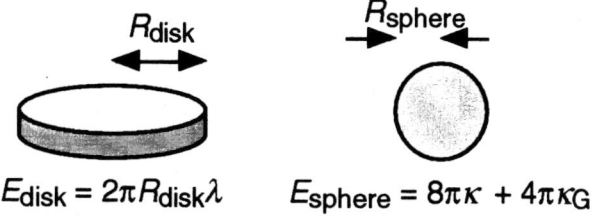

Figure 3. Energetics of disks and spheres. The two shapes have the same areas if $R_{sphere} = R_{disk}/2$.

Experiments on membrane failure find that typical bilayers rupture at tensile stresses of 1×10^{-2} J/m^2 on laboratory time scales (Needham and Hochmuth, 1989). In cells, a (two-dimensional) surface stress Π can result from the osmotic pressure difference P between the cell's interior and its external environment. For a spherical shell of radius R, the stress and pressure are related by (Fung, 1994)

$$\Pi = PR / 2. \qquad (2)$$

Thus, a spherical shell of radius 1-micron can support a pressure difference of up to 2×10^4 J/m^3, if the two-dimensional bursting stress is 1×10^{-2} J/m^2 on laboratory timescales. However, many bacteria operate at much higher internal pressures, ranging up to many atmospheres, where 1 atmosphere = 10^5 J/m^3. Most varieties of bacteria accommodate this pressure by the use of a cell wall.

Because the surface stress is proportional to R in Equation (2), a smaller cell would experience a lower stress for a given osmotic pressure P. In fact, a bilayer alone could handle an osmotic pressure of 4 atmospheres for a cell with a radius of just 50 nm, so that very small cells would not need a cell wall to function at moderate osmotic pressures. The absence of a cell wall would reduce the functional tasks of the cell and hence eliminate that part of DNA required to produce the proteins associated with cell wall construction. Alternatively, a small cell could choose to have a cell wall and increase the osmotic pressure at which it operates. Because the osmotic pressure is directly proportional to the concentration of proteins, ions, etc., then small cells could have a higher concentration of chemical reactants. Given that the rate of chemical reactions is proportional to the product of the reactant concentrations, an increase in the concentrations would result in an increase of the chemical reaction rates.

Flexible Filaments

The most evolutionarily advanced cells—eucaryotic cells—contain a filamentous cytoskeleton, which helps maintain the cell's shape, along with its other duties. Components of the cytoskeleton frequently include actin, intermediate filaments, and microtubules, with diameters in the range of 10 to 25 nm. Compared to a typical eucaryotic cell diameter of 10 microns or more, the transverse dimension of a cytoskeletal filament is trivial. Smaller cells such as bacteria, whose evolutionary origin predates eucaryotes, do not contain a cytoskeleton, but may instead possess a strong cell wall surrounding the pressurized bag bounded by a fluid membrane. Even bacteria, with a typical diameter of 1 micron, could accommodate the size of cytoskeletal filaments found in eucaryotes. However, cells with a radius as small as 50 nm would probably not have sufficient interior volume to permit a conventional cytoskeleton.

The absence of a cytoskeleton within a small cell does not imply that there are no filaments present. Cells must have some means of carrying hereditary information; the earliest cells may have used RNA but today's cells use DNA, both of which are linear molecules. Now, the visual appearance of a flexible rope, string, or linear molecule depends on the length scale of observation. For example, a human hair may be curly as seen by the eye on a length scale of centimeters, but a segment of the hair would seem straight if viewed through a microscope on a length scale of less than a millimeter. A quantity called the persistence length can be used to describe the straightness of a linear molecule. Figure 4 illustrates two linear objects; part [a] is convoluted with a short persistence length while [b] is much straighter with a long persistence length. Mathematically, the persistence length is a measure of the length scale over which a curve undergoes a significant change in direction. The arrows in Figure 4b are about a persistence length apart, as measured along the curve.

Now, double-stranded DNA has a persistence length of about 50 nm (Bustamante et al., 1994),

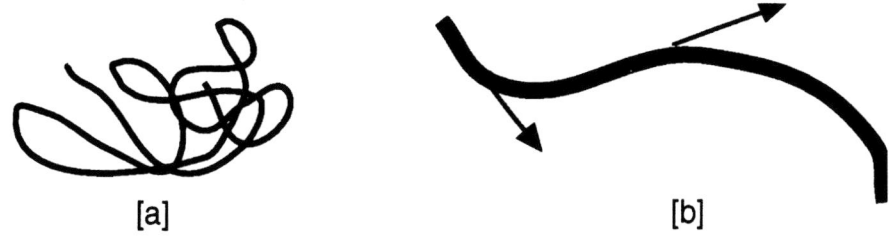

Figure 4. Schematic representation of strings with short [a] and long [b] persistence lengths. The arrows in [b] are about a persistence length apart, as measured along the contour of the string.

meaning that a 100 nm filament of DNA might look like the configuration in Figure 4b: it would appear to be neither a straight rod, nor a tangled ball of thread. At 0.34 nm per base pair, a 100 nm filament of DNA traversing the cell once would contain just 300 base pairs, not a lot of genetic information. This means that cells probably would have to be larger than 50 nm in radius to accommodate a moderate amount of DNA if it were present as a random chain. It is more likely that small cells would use RNA or another flexible molecule to carry genetic information, consistent with the idea that RNA predated DNA in evolution. Many biopolymers display a persistence length that varies as the square of the mass per unit length along the polymer, a scaling behavior consistent with the theoretical expectation that the persistence length varies as the fourth power of the radius for uniform cylindrical rods (Doi and Edwards, 1986; Landau and Lifshitz, 1986). Thus, a molecule with the same mass density as double-stranded DNA, but only half the mass per unit length, would have a persistence length of one-quarter that of DNA, just 13 nm. With a persistence length closer to 10 nm, a long molecule could be balled up in a cell of 100-nm diameter. Self-interactions along the molecule's length, as might be expected for RNA, would reduce the size of the genetic ball even further.

Acknowledgment

This work is supported in part by the Natural Sciences and Engineering Research Council of Canada.

References

1. Boal, D.H., and M. Rao. 1992. Topology changes in fluid membranes. *Phys. Rev.* **A46**: 3037-3045.
2. Bustamante, C., J.F. Marko, E.D. Siggia, and S. Smith. 1994. Entropic elasticity of λ-phage DNA. *Science* **265**: 1599-1600.
3. Doi, M., and S.F. Edwards. 1986. *The Theory of Polymer Dynamics.* Oxford: Oxford University Press, p. 316.
4. Evans, E., and W. Rawicz. 1990. Entropy-driven tension and bending elasticity in condensed-fluid membranes. *Phys. Rev. Lett.* **64**: 2094-2097.
5. Fromherz, P. 1983. Lipid-vesicle structure: size control by edge-active agents. *Chem. Phys. Lett.* **94**: 259-266.
6. Fung, Y.C. 1994. *A First Course in Continuum Mechanics.* Englewood Cliffs, New Jersey: Prentice-Hall, p. 23.
7. Helfrich, W. 1973. Elastic properties of lipid bilayers: theory and possible experiments. *Z. Naturforsch.* **28c**: 693-703.
8. Landau, L.D., and E.M. Lifshitz. 1986. *Theory of Elasticity* (3rd Ed.). Oxford: Pergamon Press, p. 67.
9. Lipowsky, R. 1991. The conformation of membranes. *Nature* **349**: 475-481.
10. Needham, D., and R.M. Hochmuth. 1989. Electromechanical permeabilization of lipid vesicles. *Biophys. J.* **55**: 1001-1009.

GENE TRANSFER AND MINIMAL GENOME SIZE

Jeffrey G. Lawrence
Department of Biological Sciences
University of Pittsburgh

Abstract

Throughout all domains of life, genetic material is exchanged within and among genomes. Horizontal transfer typically denotes rare transfer of genetic material between diverse lineages. This process does not constrain genome size in significant ways. Intraspecific recombination is more common than horizontal exchange, allows for the removal of deleterious mutations, and helps maintenance of species identity. Recombination enables organisms to maintain maximum genome sizes that are larger than those capable without gene exchange (escape of Muller's ratchet), but does not mediate potential reduction of genome size. In these cases, gene exchange allows transfer of non-essential genes among organisms, or reassortment of essential genes within a taxon. Neither process permits a cell to maintain fewer than the minimal complement of genes required for life. A model is presented whereby the frequency of gene exchange is much greater than the frequency of cell division. In this model, cells may be considered way stations for gene replication and transfer; such organisms need not maintain a full complement of genes, and genome sizes may decrease. Simulations predict the propagation of organisms where the average cell contains, on average over time, fewer than 1 gene.

Introduction

Although the influx of DNA sequence data has allowed novel approaches to assessing minimal requirements of life (1), this matter has received attention since genes and genomes were first identified (2,3). Discussion of a minimal size for a "free-living" organism necessarily includes an evaluation of that organism's genetic system. Critical components of genetic architecture include the frequency and nature of genetic exchange, that is, the transfer of genetic information among organisms. Here I will outline the nature of gene exchange mechanisms and their impact on genome size among extant organisms. Moreover, extrapolation of these mechanisms enables the elucidation of viable genetic architectures—and predictions for minimum genome sizes—for cells that are constrained in size. These models highlight potential pitfalls in the identification of organisms based on our expectations of genome content.

Modes of Gene Transfer

Horizontal Gene Transfer

Gene exchange among prokaryotic taxa is not coupled to reproduction, and may occur without direct cell-cell contact through a variety of mechanisms (e.g., transformation or bacteriophage-mediated). Since cell-cell recognition is not required for gene exchange, genetic material may be readily exchanged between distantly related lineages; this process is commonly termed lateral genetic transfer—or horizontal genetic transfer—to denote inheritance of information from outside the vertical inheritance pathway implicit in cell division (4-6). Horizontal transfer serves to shuffle genetic material among diverse organisms. Because bacterial genes are organized into operons (groups of genes whose products work together to confer a single function), horizontal transfer of DNA can result in the exchange of phenotypic capabilities among organisms, as all genes required for a particular function may be mobilized between organisms. Although this process allows for rapid adaptation of organisms to changing environments, horizontal transfer does not intrinsically limit genome size, nor does it enable substantially smaller genome sizes to be achieved (however, see selfish operons below).

Although horizontal transfer has been observed among virtually all major groups of organisms, the rate of transfer has been assessed only in *Escherichia coli* (7,8). In that lineage, the rate of introduction of DNA—16 kb/MYr—suggests that horizontal transfer plays a large role in bacterial diversification (9). However, this rate also indicates that gene transfer events occur very rarely relative to the rate of cell division.

The Selfish Operon

The organization of bacterial genes into operons likely reflects selection for mobility (10). If a function is subject to weak selection, and may be lost from an evolutionary lineage, these genes may escape evolutionary loss by horizontal transfer to a naïve host genome. Transfer is successful only if all genes required to confer a selectable phenotype are mobilized together, and all of the genes are expressed in their new host. Because the probability of cotransfer is inversely related to the distance separating the genes, those genes found in clusters will be more mobile—hence more fit—than unclustered genes. As the aggregation of the genes into a cluster does not necessarily affect the fitness of the cell, the cluster may be considered to be a selfish property of the constituent genes. Horizontal transfer serves to disseminate selfish operons among bacterial genomes, where they may confer a beneficial function to their host cells and be maintained by natural selection. This paradigm will be useful when considering models requiring very rapid gene transfer (see below).

Moreover, the process of horizontal transfer facilitates the assembly of genes into operons. As only those genes that contribute to a selectable function will be maintained after horizontal transfer, intervening genes will be removed by deletion. Cotranscription of genes will be selected because a promoter at the site of integration may accomplish expression of all genes following horizontal transfer. In this way, expression of genes in foreign hosts does not require recognition of multiple promoter sites by cells using different sets of transcriptional machinery. In a similar fashion, translational coupling will permit efficient translation of the selfish operon without need for de novo ribosome loading at each protein coding sequence. These factors warrant some small reduction of genome size as a result of horizontal transfer: the elimination of multiple promoter sites and ribosome-binding sites. However, such small sequences are minor factors when considering broad-scale reduction of genome sizes.

Intraspecific Recombination

The same mechanisms that facilitate horizontal gene exchange among distantly related bacteria also mediate intraspecific gene exchange among closely related cells. Among conspecific strains, barriers to effective recombination (e.g., differences in restriction/modification systems, and extensive DNA mispairing) are fewer, and rate of DNA exchange is greater. Intraspecific recombination among *Escherichia coli* is a common event (11), and its rate has been measured to be on the order of the mutation rate, $\sim 10^{-9}$/bp per generation (12). Estimates of the sizes of DNA fragments mediating gene exchange (0.1 to 1.0 kb) suggest that the frequency of intraspecific recombination events in *E. coli* is still lower than the rate of cell division (11).

Intraspecific recombination does directly affect genome size. A population of organisms can maintain only a finite number of genes by means of natural selection. As mutation rates (μ is some function of mutation rate) increase, fewer genes (G) can be maintained owing to selection among genes. As population size (N is some function of population size) decreases, fewer numbers of genes are maintained in the face of genetic drift. Lastly, lower rates of recombination (r is some function of recombination rate) concede the accumulation of mutations (Muller's ratchet) and enable the maintenance of fewer numbers of genes. In sum, the maximum number of genes a population can maintain can be denoted by the following relationship:

$$G \propto rN/\mu. \qquad (1)$$

Therefore, low rates of intraspecific recombination constrain the maximum number of genes a population can maintain by natural selection at any one time. Among higher eucaryotes, recombination is obligately tied to reproduction in the cycle of meiosis and syngamy. Here, the frequency of gene exchange amounts to one-half genome per generation. Although this rate is substantially higher than the rate of intraspecific recombination among prokaryotes, it serves the same purpose in affecting genome size. A population of freely recombining organisms can maintain a larger genome size, as deleterious mutations can be removed by recombination.

Empirical Approaches to Small Genomes

Equation (1) describes the influence of gene exchange on the maintenance of genes by natural selection. Notably, recombination facilitates the removal of deleterious alleles (13) and allows for the simultaneous maintenance of larger numbers of genes by natural selection. Implicit in this discussion—and in the concept of selfish operons—is the idea that many genes found in bacterial cells are not essential for survival. Indeed, the genomes of every organism tested include genes that are not essential for life. Although the *E. coli* genome bears over 4,500 genes (14), surveys of conditional mutations reveal that fewer than 10% of these loci are essential. Even the *Mycoplasma genitalium* genome—at 580073 bp and ~470 genes, which represents the smallest bacterial genome to date (15)—is not composed completely of essential genes (16,17). Comparisons among sequenced genomes suggest that only 256 genes may be required to support a recognizable bacterial cell (1).

These approaches can help describe a minimal gene set enabling the growth of a prokaryotic bacterium, but they are constrained to encoding a sophisticated, highly evolved set of inter-dependent biochemical reactions. These exercises preclude, for example, the definition of a minimal set of genes based on self-replicating ribozymes. From an exploratory perspective, these approaches do not encompass the definition of potential sets of minimal genes that exploit alternative biochemistries. Therefore, we must divorce discussion of the role of genetic transfer on minimal genome size from the preconceptions of cellular biology. Below, I will develop a context-independent model describing how rapid gene transfer predicts the maintenance of very small genome sizes when cells are constrained to small sizes.

Model of Minimal Genome Size

Minimum Genome Composition and the Cellular Environment

What is represented by the smallest collection of 256 essential genes described by Mushegian and Koonin (1), even for a biochemically complex organism like *Mycoplasma*? It is the group of genes that define and describe the cellular environment, in which all genes are replicated. This collection of genes comprises a mutually reliant group; without the function of any one of the genes, the cell cannot survive. More specifically, without the functions of any one of these genes, none of the constituent genes can replicate. In this way, one may consider the cell to be an environment in which genes can replicate. The minimum subset of genes whose products define the cell describes a group with an emergent property: the ability to control their own environment. Regardless of what functions one requires a minimal cell to perform, some subset of replicating genes must be working together to maintain this environment; outside of this environment, genes replicate very poorly. The products of this minimal subset of genes modify the environment so that the group may replicate more efficiently, thereby increasing their fitness. We will call this group of genes the *cellular consortium.*

Horizontal transfer describes the transfer of genes that do not belong to the cellular consortium; rather, these genes may increase the fitness of the consortium in certain environments (like the *lac* operon aids *E. coli* growth), but these selfish operons are not required for cell growth. Intraspecific recombination describes mechanisms of reassorting members of the cellular consortium, but does not allow reduction of this group below the minimal number of genes required to perform cellular function. For gene exchange to affect this minimal number of genes, whose number and nature depend entirely on the properties of the minimal cell, we must speculate how constraints on cell size permit *fewer* than this minimal number of genes to be present in a cell at any one time. To do this, we must model the assembly of the cellular consortium, and devise a mechanism whereby constraint on cell size permits cells to replicate with fewer than the minimal number of genes required to form the cellular consortium.

Single Replicon Dynamics and the Cellular Consortium

To begin this model, we will consider the self-replicating gene—outside the context of the cell—to be the smallest living creature. A molecule that can self-replicate will increase in numbers as it consumes available resources. Consider the distribution of resources to be non-uniform, where the preferred environments are micelles (to use a familiar term). A successful self-replicating gene would travel from micelle to micelle, consuming the resources contained therein and replicating its genome. At this point, the cellular consortium does not exist, and the self-replicating gene leads a nomadic existence, traveling from resource patch to resource patch to replicate.

Mutants may arise among the self-replicating genes that enable greater replication in the micelle environments; such mutants could, for example, perform some simple biochemical functions that replenish some portion of the available nutrient pool. Different mutants may arise, each performing some different biochemical feat that enables it to replicate to a greater degree in some environment. We may consider these different classes of replicons as protospecies, each of which can replicate successfully in a different set of micelle environments; a hypothetical collection of genes, each with an elementary function, is listed in Table 1. At this stage, the replicons exploit the micelles as resource patches, traveling from micelle to micelle to replicate.

Such a system is shown in Figure 1. Here, each of the four genes listed in Table 1 is rapidly transferred between micelles. If a gene enters a micelle bearing all nutrients required for replication except one, and the replicon encodes a function allowing the synthesis of that compound, the gene and the micelle can replicate (replication-competent micelles). Following division, each daughter micelle would bear high concentrations only of the compound synthesized by the resident gene. For micelle and gene division to occur again, the micelle must be visited by each of the other three replicons.

If the replicons encoding the different functions assemble into a consortium, the most fit consortia would be that which combines a set of functions that would allow exploitation of the largest number of

Table 1 Participants in Micelle Simulation

Participant	Function[a]
Gene 1	Synthesizes compound A; replicates when provided with compounds B, C, and D
Gene 2	Synthesizes compound B; replicates when provided with compounds A, C, and D
Gene 3	Synthesizes compound C; replicates when provided with compounds A, B, and D
Gene 4	Synthesizes compound D; replicates when provided with compounds A, B, and C
Compound A	Required for gene replication; synthesized from precursors by Gene 1
Compound B	Required for gene replication; synthesized from precursors by Gene 2
Compound C	Required for gene replication; synthesized from precursors by Gene 3
Compound D	Required for gene replication; synthesized from precursors by Gene 4
Micelle	Enclosed environment maintaining compounds A, B, C, and D

[a]Function in computer simulation (Figure 1).

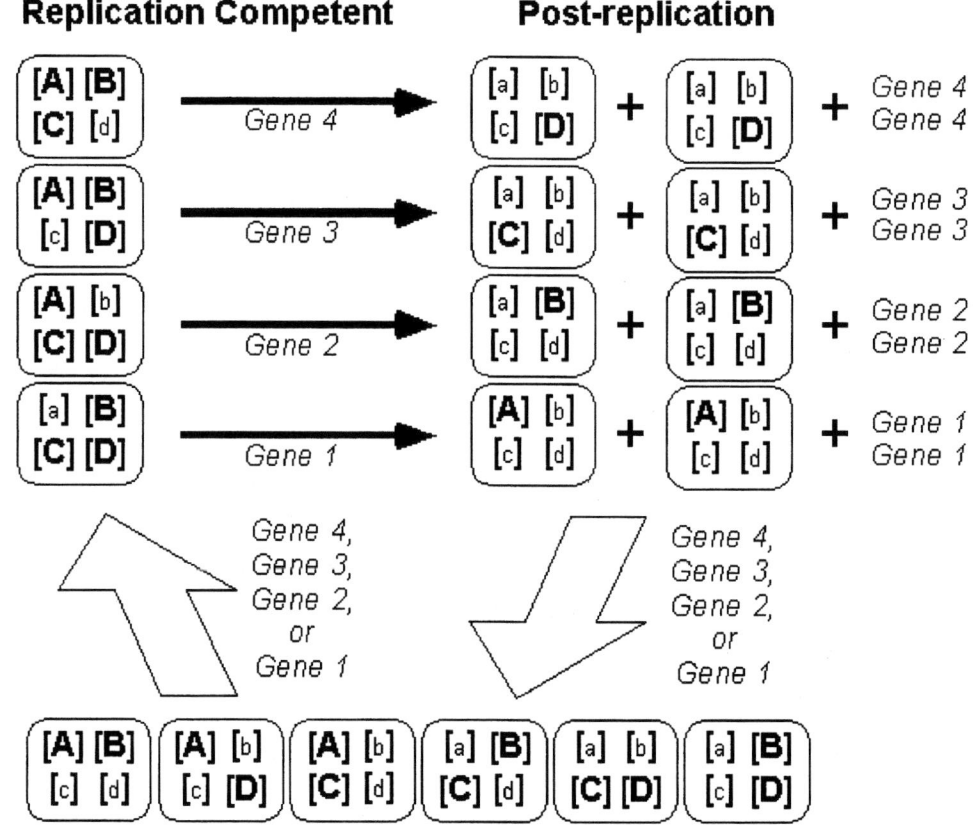

Figure 1. Model for micelle propagation of a meta cell. Enclosed boxes represent micelles, which require four compounds to enable gene and micelle division (Table 1). The [X] symbols represent sufficiently high concentrations of compound X to allow replication; the [x] symbols represent post-replication levels.

environments (genes 1, 2, 3, and 4 on a single segment). If cell growth, i.e., propagation of the micelle, provides for better replication of the cellular consortium than does the nomadic lifestyle, a critical transition would occur. The genes comprising the cellular consortium will abandon their nomadic lifestyle and take up agriculture, using their battery of biochemical functions to maintain their preferred micelle environment. This cellular consortium bears the emergent property of cell growth, that is, the propagation of the preferred environment.

Rapid Gene Transfer of the Cellular Consortium and the Meta-Cell

Computer simulations of this model show rapid coalescence of replicons to form a cellular consortium that outcompetes non-cellular, nomadic replicons for available nutrients. However, a tacit assumption of the model is that all necessary pieces—all the individual genes—can coexist in the same cell. If individual micelles cannot support the entire subset of genes required to form the cellular consortium, a "cell" containing the entire cellular consortium cannot evolve. That is, the collection of cooperating genes cannot abandon their nomadic lifestyle in favor of metabolic agriculture. Rather, mini-consortia of replicons maintain their nomadic existence, traveling from micelle to micelle to replicate. At each stop, they may replicate using some subset of the available nutrients it requires, while replenishing others. This collection of mini-consortia replicates among these way stations of micelles. Here, the cell never forms by assembly of the cellular consortia because of constraints on cell size. Rather, the meta-cell develops as subsets of the cellular consortia perform their functions in a temporally and spatially diffuse manner. If propagation of the micelle environment is more favorable than a nomadic lifestyle, as suggested above for the evolution of the cell, the cellular consortium will evolve to propagate micelle. As each micelle cannot house all members of the cellular consortium at the same time, each member of the mini-consortia must travel through any one micelle to allow for micelle division. Therefore, in this model, the rate of transfer of genes among micelles is much more frequent than is micelle division. Rapid gene exchange allows for propagation of the meta-cell organism.

The Specter of Group Selection

Group selection is a framework for understanding cooperativity among competing organisms; each member makes a contribution to the group, and all members benefit. Group selection models are unstable in that members of the group can "cheat" by extracting the benefits of the group without making a contribution. The cellular consortium model does not require group selection for maintenance. Each mini-consortium of genes must perform some function to maintain the meta-cell. Cheaters that perform no biochemical function cannot replicate, as there is no member of the meta-cell that contains a complete complement of nutrients; only the mini-consortia that perform the required function can replicate there. Cheaters could arise as mutants of a mini-consortium that consume the nutrients and fail to synthesize enough of its product to feed other members (who will visit the micelle at a later time) that lack this function. Meta-cells are susceptible to this mutant, but they are less fit, because the micelles lacking this key nutrient accumulate and the meta-cell dies. Simulations show that natural selection maintains meta-cells with a minimal complement of non-contributing members.

Perspective on the Minimum Genome Size

The meta-cell model is a stable means of propagating replicons using high rates of gene transfer among micelles that cannot support the entirety of the cellular consortium. In this case, the rate of transfer of genes among micelles is far more rapid than the rate of micelle division. Such systems are stable if size constraints prevent the assembly of the cellular consortium in a single micelle to form a cell. In this model, information-bearing micelles—those containing a mini-consortium at any one time—may contain as few as 1 gene. Since all micelles in the meta-cell do not contain genes at all times (each mini-consortium must exit a micelle to allow entry of another mini-consortium), the average genome size may be less than one gene. One may consider the meta-cell to be a single-celled organism whose genome is distributed through a network of micelles. If rapid transfer of genetic material defines a genetic architecture, a cell is not limited to containing all of the genes required for growth.

Acknowledgments

I thank Drs. Anthony Bledsoe, Susan Kalisz, and Roger Hendrix for helpful discussions. This work was supported by grants from the Alfred P. Sloan Foundation and the David and Lucile Packard Foundation.

References

1. Mushegian, A.R., Koonin, E.V. (1996). *Proc. Natl. Acad. Sci., USA* **93**, 10268-10273.
2. Pirie, N.W. (1973). *Ann. Rev. Microbiol.* **27**, 119-132.
3. Haldane, J.B.S. (1928). *Possible Worlds and Other Papers* (Harper and Brothers, New York).
4. Syvanen, M., Kado, C.I. (1998). *Horizontal Gene Transfer* (Chapman and Hall, London).
5. Syvanen, M. (1994). *Ann. Rev. Genet.* **28**, 237-261.
6. Kidwell, M. (1993). *Ann. Rev. Genet.* **27**, 235-256.
7. Lawrence, J.G., Ochman, H. (1998). *Proc. Natl. Acad. Sci., USA* **95**, 9413-9417.
8. Lawrence, J.G., Ochman, H. (1997). *J. Mol. Evol.* **44**, 383-397.
9. Lawrence, J.G. (1997). *Trends Microbiol.* **5**, 355-359.
10. Lawrence, J.G., Roth, J.R. (1996). *Genetics* **143**, 1843-1860.
11. Milkman, R. (1997). *Genetics* **146**, 745-750.
12. Guttman, D.S., Dykhuizen, D.E. (1994). *Science* **266**, 1380-1383.
13. Muller, H. (1932). *Am. Nat.* **66**, 118-138.
14. Blattner, F.R., Plunkett, G.R., Bloch, C.A., Perna, N.T., Burland, V., et al. (1997). *Science* **277**, 1453-1474.
15. Fraser, C.M., Gocayne, J.D., White, O., Adams, M.D., Clayton, R.A., et al. (1995). *Science* **270**, 397-403.
16. Arigoni, F., Talabot, F., Peitsch, M., Edgerton, M.D., Meldrum, E., et al. (1998). *Nat. Biotechnol.* **16**, 851-856.
17. Razin, S. (1997). *Indian J. Biochem. Biophys.* **34**, 124-130.

Panel 2

Is there a relationship between minimum cell size and environment?
Is there a continuum of size and complexity that links conventional bacteria to viruses?
What is the phylogenetic distribution of very small bacteria?

DISCUSSION

Summarized by Kenneth Nealson, Panel Moderator

Goals of the Session

Panel 2 focused in general discussions on the issue of whether any given kind of environment appeared to favor very small microbes. Experts with experience in a wide variety of intracellular and extracellular niches, including host cells (Van Etten and Kajander), aquatic environments (Button and DeLong), hydrothermal environments (Stetter and Adams), and soils and sediments (Staley) presented their views (Table 1). Insofar as it was possible, the discussion was focused on questions relating to the size ranges of organisms found in each environment, and the question of whether some properties of the environment (nutritional, physical, or chemical) might lead to the favoring of very small, nanometer-sized cells. In essence, this discussion sought to use the natural experiences of field and laboratory microbiologists to reach consensus on questions such as the following:

1. What are the smallest sizes of viable organisms actually seen in the environment?
2. What are the environmental issues that impose or relieve restrictions on cell size?
3. What strategies are used to attain and maintain small size in nature?

Organisms Encountered in Natural Environments

What are the smallest viable organisms actually encountered in the various environments? In pursuit of the answer to this question, the speakers focused on their environments of interest (see Table 1) and the sizes of organisms encountered there. Included were organisms such as obligate parasites and symbionts, as well as free-living organisms, both rapidly growing and in various types of resting stages. For the sake of completeness, mitochondria and chloroplasts were included, although no

Table 1 Organisms, Environments, and Presenters

Organism	Environment	Speaker
Viruses	Animal or plant cells	Van Etten
Nanobacteria	Animal serum	Kajander
Attached bacteria	Soils, sediments, rocks	Staley
Hyperthermophiles	Hot springs and vents	Stetter
Hyperthermophiles	Hot springs and vents	Adams
Aquatic bacteria	Lakes and oceans	Button; DeLong

Table 2 Size Ranges of Organisms or Organelles, and Niches Where They Are Found

Organism	Diameter Range (nm)	Life Style
Virus	30 to 200	Host-dependent
Nanobacteria	100 to 200	Host-dependent
Marine bacteria	100 and larger	Free-living
Attached forms	100 and larger	Free-living
Hyperthermophiles	200 and larger	Free-living
Mitochondria	200 and larger	Host-dependent
Chloroplasts	200 and larger	Host-dependent

presentations were specifically made in these areas. The size ranges shown in Table 2 represent the consensus values reached in the presentations and in ensuing discussions by the assembled group. In many cases it was hard to reach consensus on firm estimates for the smallest organisms or organelles encountered, and the reader is referred to specific arguments in the individual papers. For example, there was considerable debate with regard to the nanobacteria, as summarized by Dr. Kajander. While such nanobacteria have been reported to be smaller than 100 nm in diameter, Dr. Kajander was of the opinion that the only organisms for which growth could be established with certainty were those of 100-nm diameter or larger. This represents an area of considerable importance in terms of being able to search for and recognize very small organisms (e.g., Are there organismal fragments that appear to have similar morphologies, but are not actually viable, growing entities?).

A point of interest with regard to this area is that virtually all of the microbiologists present had encountered structures resembling cells in the size range of 100 to 200 nm, but whether or not these could be demonstrated to be viable or cultivable microbes had usually not been established. The timeworn method of filtration through a 200-nm (0.2 micrometer) pore-size filter was still very dependable in terms of delineating cultivable bacteria.

Environmental Parameters and Size

What are the environmental issues that may impose or relieve restrictions on the smallest sizes that can be achieved by organisms? In pursuit of this question, the speakers considered a variety of different environmental factors that might lead to organisms adopting a smaller size. These included:

1. Nutrient-rich environments, which allow evolution to small cells with less biosynthetic capacity, such as obligate parasites or symbionts;
2. Nutrient-poor environments, which lead to adaptation of small, starved cells;
3. High or low temperature; and
4. Attachment to surfaces.

Of the issues discussed, that of nutrient availability was repeatedly noted as one of potential importance. Two major issues were emphasized: (1) the effect of nutrient limitation and starvation, which leads to adaptation of normally large cells to resting stages that are considerably smaller; and (2) the effect of nutrient richness, which leads to evolution of cells that are host dependent, and often considerably smaller.

In nutrient-poor environments, organisms were deemed to be small in the starved state, although the lower size limit of this starvation state appears to be on the order of 200 nm. The mechanisms for achieving such small size (or for returning to a state of larger, rapidly growing cells) are not well understood. However, such organisms are not regarded as true nanobacteria, because under nominal growth conditions, they are considerably larger than the diminutive forms discussed here. These larger forms are thought to represent a true evolutionary lower size limit for DNA-based life.

In the case of intracellular symbiosis or parasitism in nutrient-rich environments, considerable discussion occurred as to whether or not such organisms could eliminate enough functions to evolve to a very small size. Dr. Adams presented a general discussion of the theoretical limits of life, based on organisms with the same basic biochemistry as those we are familiar with. At the theoretical extreme are the viruses, which are obligate intracellular parasites and which have no need for their own transport systems, translation machinery, or transcription apparatus. These organisms can be quite small, as they consist of a protein coat surrounding the genetic material. The lower size limits are seen in some RNA viruses like the Qβ virus (which contains only three genes), and in certain animal viruses (e.g., poliovirus) that are in the range of 25 to 50 nm in diameter, while most others are in the range of 100 to 200 nm or even larger. Symbiotic organelles or bacteria are also commonly found in the 200-nm range and are sometimes smaller. These include non-cultivable bacteria from a wide variety of organisms, intracellular organelles (e.g., mitochondria or chloroplasts), and the enigmatic nanobacteria discussed by Dr. Kajander.

It should be clear, however, that the strategies used for attaining and maintaining small size will be very different for the oligotrophic organisms, which become small as a matter of optimizing their surface-to-volume ratio under diffusion-limited growth conditions, and the eutrophic organisms, which are allowed to become small because of the richness of their environment. In the latter case, these organisms are not faced with the maintenance of the genetic or physiological capacity for either extensive biosynthesis or diverse catabolism. While it is often possible to maintain such "obligate" symbionts or parasites in a host-free growth phase using a very rich medium, discussion of their role(s) as very small bacteria may be relevant only in the context of their existence as parasites or symbionts.

Perhaps the liveliest discussion in Panel 2 centered on the specification of the smallest sizes actually seen in the environment and the criteria that one accepts for a living cell. To this end, Dr. Kajander proposed that nanobacteria may fragment into non-growing entities that appear considerably smaller than the true, viable organisms, and that these fragments may come together at a later time to form a viable organism. In terms of this possibility, Dr. Van Etten pointed out that some plant viruses exhibit just such a pattern. Each particle packages separate RNA, and sometimes three separate particles are needed to establish an infection. It was also noted that many estimates of the smallest sizes for viable organisms come from filtration studies, and that bacteria with non-rigid cell walls may pass through filters of pore size smaller than their actual diameter.

As a final point, one would like to have an indication of the minimum cell volume needed to sustain life. Dr. de Duve emphasized that diameter alone is not a sufficient parameter, pointing out the practical difficulty of estimating true diameter from random thin sections To this end, the discussion by Dr. Adams focused almost entirely on the intracellular volumes of variously sized and shaped organisms, and the possibility that such volumes could accommodate the machinery of life.

Strategies for Attaining and Maintaining Small Cell Size

Are there strategies that can allow the minimum size of an organism to be smaller than might be anticipated through studies of extant organisms? With regard to this question, several strategies were considered by Panel 2 speakers. The first, discussed briefly above, was that of Kajander and Van Etten, in which organisms actually fragment so that each very small organism is incapable of growth, but the population is capable of achieving success. While this strategy is known for some RNA viruses, there are as yet no examples among the prokaryotes.

A second strategy considered was that employed by parasites and symbionts, which simply discard a sizable fraction of their genetic information and adopt a host-dependent life style. Such organisms, while achieving a very small size, sacrifice the freedom of being host-free.

Other approaches that might allow attainment and maintenance of a smaller cell size are (1) reduction of the average size of proteins; (2) an RNA-world approach in which a single type of molecule accomplishes both catalytic and genetic functions; and (3) the use of overlapping genes and genes on complementary strands. In no case has a systematic analysis of any of these approaches been done.

Consensus?

In terms of reaching a consensus, Panel 2 members, with the exception of Dr. Kajander, who described nanobacteria in the size range of 100 nm, considered that the lower size limit of bacteria-like particles believed to be cultivable corresponded to spherical organisms with a diameter in the size range of 200 to 250 nm. The nanobacteria of Kajander are "obligate" parasites (e.g., they require very rich media to achieve host-free growth) and so may fall into the category of organisms adopting a host-dependent life style. Thus, despite a very large amount of discussion, a general consensus was reached that was in agreement with the theoretical arguments put forward during the workshop, that the lower limit of size for a free-living, DNA-based organism corresponds to a spherical organism with a diameter in the size range of 200 to 250 nm. For host-dependent organisms the size may be smaller, and the extent of the smallness will certainly depend on the extent to which genetic and physiological functions have been discarded.

For an organism that used one type of molecule for both catalysis and replication, the size could be considerably smaller, as discussed by Dr. Benner and others.

CAN LARGE dsDNA-CONTAINING VIRUSES PROVIDE INFORMATION ABOUT THE MINIMAL GENOME SIZE REQUIRED TO SUPPORT LIFE?

James L. Van Etten
Department of Plant Pathology
University of Nebraska at Lincoln

Abstract

The genomes of a few viruses, such as *Bacillus megaterium* phage G (670 kb) and the chlorella viruses (330 to 380 kb), are larger than the predicted minimal genome size required to support life (ca. 320 kb). A comparison of the 256 proteins predicted to be required for life with the putative 376 proteins encoded by chlorella virus PBCV-1, as well as those encoded by other large viruses, indicates that viruses lack many of these "essential" genes. Consequently, it is unlikely that viruses will aid in determining the minimal number and types of genes required for life. However, viruses may provide information on the minimal genome size required for life because the average size of genes from some viruses is smaller than those from free-living organisms. This smaller gene size is the result of three characteristics of virus genes: (1) virus genes usually have little intragenic space between them or, in some cases, genes overlap; (2) some virus-encoded enzymes are smaller than their counterparts from free-living organisms; and (3) introns occur rarely, if at all, in some viruses.

Introduction

Two recent estimates of the minimum genome size required to support life arrived at similar values. (1) The effect of 79 random mutations on the colony-forming ability of *Bacillus subtilis* resulted in the conclusion that a genome of 318 kb could support life (Itaya, 1995). Assuming 1.25 kb of DNA per gene (Fraser et al., 1995), this amount of DNA would encode 254 proteins. (2) A comparison of the genes encoded by *Mycoplasma genitalium* and *Haemophilus influenzae* led Mushegian and Koonin (1996) to suggest that as few as 256 genes are necessary for life. Using the same 1.25 kb gene size, the minimum self-sufficient life-form would have a 320 kb genome. Interestingly, these estimates are smaller than the genomes of some viruses (Table 1). Bacteriophage G, which infects *Bacillus megaterium*, has a genome of about 670 kb (Hutson et al., 1995); phycodnaviruses that infect chlorella-like green algae have 330 to 380 kb genomes (Rohozinski et al., 1989; Yamada et al., 1991); and some insect poxviruses have genomes as large as 365 kb (Langridge and Roberts, 1977). Other large, dsDNA-containing viruses, such as herpesviruses, African swine fever virus (ASFV), coliophage T4, baculoviruses, and iridoviruses, have genomes ranging from 100 to 235 kb (see Table 1). However, except for the common property of having large dsDNA genomes, these viruses differ significantly from one another in such characteristics as particle morphology, genome structure, and the intracellular site of replication. For example, poxviruses, herpesviruses, and baculoviruses have an external lipid envelope, whereas iridoviruses and phycodnaviruses have an internal lipid component. Baculovirus genomes are circular, iridoviruses and phage T4 have linear circular permuted genomes with terminal redundancy, and the linear genomes of herpesviruses have sequences from both termini that are repeated internally in an inverted form. The phycodnaviruses, poxviruses, and ASFV have linear genomes with covalently closed hairpin ends. Finally, herpesviruses and baculoviruses primarily replicate in the nucleus, whereas

Table 1 Representative Large dsDNA Viruses

Virus[a]	Virus Family	Host	Genome Size (bp)	Minimum No. of Codons[b]	No. of Genes	Average Length of Gene (Bases)	Reference
Phage G	Myoviridae	*Bacillus megaterium*	~670,000	—	—	—	Hutson et al., 1995
PBCV-1	Phycodnaviridae	Chlorella NC64A	330,742	65	376[c]	880	Li et al., 1997
MsEPV	Poxviridae	Grasshopper	236,120 (222,120)[d]	60	267 (257)[d]	884 (864)[d]	Afonso et al., 1998
MCV	Poxviridae	Human	190,289 (180,889)[e]	60	182 (180)[e]	1,046 (1,005)[e]	Senkevich et al., 1996
ASFV	Unclassified	Swine	170,101 (166,613)[f]	60	151	1,127 (1,103)[f]	Yanez et al., 1995
Coliphage T4	Myoviridae	*E. coli*	168,800	29	288[g]	586	Kutter et al., 1994
HSV-2	Herpesviridae	Human	154,746	—	74[h]	2,091	Dolan et al., 1998
AcNPV	Baculoviridae	Insects	133,894	50	154	890	Ayres et al., 1994
LCDV	Iridoviridae	Flounder	102,653	40	110	933	Tidona and Darai, 1997

[a]G, Giant; PBCV-1, *Paramecium bursaria chlorella* virus 1; MsEPV, *Melanoplus sanguinipes* entomopoxvirus; MCV, *Molluscum contagiosum* virus; ASFV, African swine fever virus; HSV-2, Herpes simplex virus type 2; AcNPV, *Autographa californica* multinucleocapsid nuclear polyhedroses virus; LCDV, lymphocystis disease virus.
[b]Minimum number of codons used by the authors to calculate an open reading frame (ORF).
[c]Four of the genes are diploid.
[d]MsEPV has a 7 kb inverted repeat at each terminus. This 14 kb encodes 10 small ORFs (60 to 155 codons). Removal of 14 kb and 10 ORFs from the calculations produces the smaller genome size (in parentheses).
[e]MCV has a 4.7 kb inverted repeat at each terminus. This 9.4 kb encodes two 488 codon ORFs. Removal of 9.4 kb and 2 ORFs from the calculations produces the smaller genome size (in parentheses).
[f]ASFV has a 2134 bp inverted repeat at each terminus. The most terminal 1744 bp at each end do not encode an ORF and thus 3488 bp were removed from the calculations, which leads to the smaller genome size (in parentheses).
[g]This includes 161 genes known to encode proteins and 127 suspected of encoding proteins (Gisela Mosig, personal communication).
[h]HSV-2 has 473 met-initiated ORFs of 50 codons or longer of which 74 are known to be functional genes. If some of the additional 399 ORFs prove to be protein encoding, the average length of a herpesvirus gene would decrease substantially.

the entire life cycle of the poxviruses occurs in the cytoplasm. Iridoviruses and phycodnaviruses initiate replication in the nucleus, but capsids are assembled and DNA is packaged in the cytoplasm.

With the exception of bacteriophage G, the genome of at least one representative of each of these dsDNA-containing viruses has been sequenced, and the number of putative genes encoded by the viruses are listed in Table 1. Because the 330,742 bp genome of the phycodnavirus PBCV-1 is the largest virus genome sequenced to date (Lu et al., 1995, 1996; Li et al., 1995, 1997; Kutish et al., 1996), it will be used to illustrate the organization and diversity of genes that can be encoded by a large dsDNA-containing virus. The PBCV-1 genome encodes 701 open reading frames (ORFs), defined as continuous stretches of DNA that translate into a polypeptide initiated by an ATG translation start codon, and extending 65 or more codons. The 701 ORFs have been divided into 376, mostly non-overlapping, ORFs (major ORFs), which are predicted to encode proteins, and 325 short ORFs, which are probably non-protein encoding. Four PBCV-1 ORFs reside in the 2.2 kb inverted terminal repeat region of the PBCV-1 genome and consequently are present twice in the PBCV-1 genome (Strasser et al., 1991; Lu et al., 1995). The 376 PBCV-1 ORFs are evenly distributed along the genome and, with one exception, there is little intergenic space between them. The exception is a 1788-bp non-protein coding sequence near the center of the genome. This region, which has numerous stop codons in all reading frames, does code for ten tRNA genes. The middle 900 bp of this intergenic region also has some characteristics of a "CpG island" (Antequera and Bird, 1993). To put the coding capacity of the PBCV-1 genome in perspective, the 580-kb genome of the smallest self-replicating organism, *Mycoplasma genitalium* encodes about 470 genes (Fraser et al., 1995).

Computer analyses of the predicted products of the 376 PBCV-1 major ORFs indicate that about 40% of the ORFs resemble proteins in the databases, including many interesting and unexpected proteins. Some PBCV-1 encoded proteins resemble those of bacteria and phages, such as DNA restriction endonucleases and methyltransferases. However, other PBCV-1 encoded proteins resemble those of eukaryotic organisms and their viruses, such as translocation elongation factor-3, RNA guanyltransferase, and two proliferating cell nuclear antigens. The PBCV-1 genome is thus a mosaic of prokaryotic- and eukaryotic-like genes, suggesting considerable gene exchange in nature during the evolution of these viruses.

This gene diversity undoubtedly reflects the natural history of the chlorella viruses. The viruses are ubiquitous in freshwater collected worldwide, and titers as high as 4×10^4 infectious viruses per ml of native water have been obtained (Van Etten et al., 1985; Yamada et al., 1991). The only known hosts for these viruses are chlorella-like green algae, which normally live as hereditary endosymbionts in some isolates of the ciliate, *Paramecium bursaria*. In the symbiotic unit, algae are enclosed individually in perialgal vacuoles and are surrounded by a host-derived membrane (Reisser, 1992). The endosymbiotic chlorella are resistant to virus infection and are only infected when they are outside the paramecium (Van Etten et al., 1991).

Because of the large size of the PBCV-1 genome, it is not surprising that many of the predicted 376 PBCV-1 genes have not been found in other viral genomes. Box 1 lists some of the PBCV-1 encoded ORFs that match proteins in the databases and, in a few cases, indicate if a gene is transcribed early (E) or late (L) during virus replication. The functionality of some PBCV-1 encoded proteins has been established by either complementation of mutants and/or assaying recombinant protein for enzyme activity. (These proteins are indicated with an asterisk in Box 1.) Twenty-nine of the PBCV-1 ORFs resemble one or more other PBCV-1 ORFs suggesting that they might be either gene families or gene duplications. Sixteen families have 2 members, 8 families have 3 members, 3 families have 6 members, and 2 families have 8 members.

Even if some of the suspected 376 PBCV-1 protein-encoding genes turn out to be non-coding, it is clear that PBCV-1 encodes more genes than the minimum number predicted to be necessary to support life. Comparing the genes that Mushegian and Koonin (1996) proposed were essential to support life with the PBCV-1 encoded genes indicates that the virus lacks many of these genes, including a RNA polymerase, a complete protein synthesizing system, and an energy-generating system. Consequently, PBCV-1 depends on the algal host to fulfill these essential functions.

A comparison of the genes encoded by the other large dsDNA-containing viruses listed in Table 1 with those encoded by PBCV-1 indicates that a few genes are present in all of the viruses, e.g., each of the viruses encodes a DNA polymerase gene. However, there are more differences in the genes encoded by these viruses than similarities, which reflects the different life-styles of the viruses. Like PBCV-1,

Box 1 Putative ORFs Encoded by Chlorella Virus PBCV-1[a]

DNA Replication & Repair
- (E) A185R DNA polymerase
- (E) A544R* DNA ligase
- (E) A583L* DNA topoisomerase II
- A193L PCNA
- A574L PCNA
- A153R Helicase
- A241R Helicase
- A548L Helicase
- (E) A50L* T4 endonuclease V
- A39L CyclinA/cdk associated protein
- A638R Endonuclease

Nucleotide Metabolism
- (E) A169R* Aspartate transcarbamylase
- A476R Ribo. reductase (small subunit)
- A629R Ribo. reductase (large subunit)
- A427L Thioredoxin
- A438L Glutaredoxin
- A551L dUTP pyrophosphatase
- A596R dCMP deaminase
- A416R dG/dA kinase
- A363R Phosphohydrolase
- A392R ATPase
- A674R *Dicty* Thy protein

Transcription
- A107L RNA transcription factor TFIIB
- A125L RNA transcription factor TFIIS
- A166R Exonuclease
- A422R Endonuclease
- (E) A103R* RNA guanyltransferase
- A464R RNase III

Translation
- (E,L) A666L Translation elongation factor-3
- A85R Prolyl 4-hydroxylase alpha-subunit
- A105L Ubiquitin C-terminal hydrolase
- A448L Protein disulphide isomerase
- A623L Ubiquitin-like fusion protein
- 10 tRNAs

Cell Wall Degrading
- (E) A181R* Chitinase
- (L) A260R* Endochitinase
- (L) A292L* Chitosanase
- A94L β-1,3 glucanase

DNA Restriction/Modification
- A251R* Adenine DNA methylase (M.CviAII)
- A252R* Restriction endonuclease (R.CviAII)
- A252R* Restriction endonuclease (R.CviAII)
- (E) A517L* Cytosine DNA methylase (M.CviAIII)
- (L) A530R* Cytosine DNA methylase (M.CviAIV)
- (E) A581R* Adenine DNA methylase (M.CviAI)
- (E) A579L* Restriction endonuclease (R.CviAI)
- A683L Cytosine DNA methylase (M.CviAV)

Sugar and Lipid Manipulation
- (L) A64R Galactosyl transferase
- (E) A98R* Hyaluronan synthase
- (E) A100R* Glucosamine synthase
- A114R Fucosyltransferase
- (E) A118R GDP-D-mannose dehydratase
- A222R Cellulose synthase
- A295L Fucose synthase
- (E) A473L Cellulose synthase
- (E) A609L* UDP-glucose dehydrogenase
- A49L Glycerophosphoryl diesterase
- A53R 2-hydroxyacid dehydrogenase
- A271L Lysophospholipase

Phosphorylation/dephosphorylation
- A34R Protein kinase
- (L) A248R* Phosphorylase B kinase
- A277L Ser/Thr protein kinase
- A278L Ser/Thr protein kinase
- A282L Ser/Thr protein kinase
- A289L Ser/Thr protein kinase
- A305L Tyr phosphatase
- A614L Protein kinase
- A617R Tyr-protein kinase

Miscellaneous
- A207R* Ornithine decarboxylase
- A217L Monoamine oxidase
- (L) A237R* Homospermidine synthase
- A78R β-alanine synthase
- A245R Cu/Zn-superoxide dismutase
- A284L* Amidase
- A465R Yeast ERVI protein
- A598L Histidine decarboxylase
- A250R K+ ion channel protein
- A625R Transposase

[a] E and L refer to early and late genes, respectively. An asterisk means that the gene encodes a functional enzyme as determined either by complementation or by enzyme activity of a recombinant protein.

each of these viruses rely on their host cells for such basic functions as energy generation, protein synthesis, and amino acid biosynthesis. The net result is that it seems unlikely that examining virus genes will aid in determining the minimal number and types of genes required to support life.

On the other hand, viruses may provide useful information about the minimum genome size required for the genes to support life. In Table 1, we have calculated the average length of a virus gene by dividing the genome size by the number of putative genes. Except for herpesvirus HSV 2, the size of the average virus gene varied from 586 nucleotides for coliphage T4 to 1,127 nucleotides for ASFV, with the average gene size for five of the viruses being less than 1 kb. The sizes are even smaller if one removes the non- or sparsely-coding regions in the virus genomes before making the calculations. For example, the two poxviruses MsEPV and MCV, as well as ASFV, have inverted terminal repeat regions that either are non-coding or only encode a few genes. Eliminating these non-coding regions from the calculations reduces the size of the average MsEPV gene from 884 nucleotides to 864 nucleotides, the MCV gene from 1,046 nucleotides to 1,005 nucleotides and ASFV from 1,127 nucleotides to 1,103 nucleotides (Table 1).

Similar calculations made on nine Eubacteria and three Archaea indicate that the average length of Eubacteria protein-encoding genes ranges from 1,023 nucleotides for *Aquifex aeolicus* to 1,234 nucleotides for *Mycoplasma genitalium* (Doolittle, 1998). The predicted average length of the three archaea is slightly smaller—895, 943, and 961 nucleotides for *Archaeoglobus fulgidus, Methanococcus thermoautotrophicum,* and *M. jannaschii*, respectively. Thus, depending on the virus and bacterium being compared, the average functional virus gene is 10 to 50% smaller than the average bacterial gene. This conclusion depends on the assumption that at least the majority of the predicted virus genes, in fact, encode proteins.

The apparent smaller size of genes from large dsDNA viruses can be attributed to three factors. (1) Typically, virus genomes have little intergenic space and, in some cases, genes overlap. This tight packaging of genes does not prevent gene regulation, however, as virus genes are typically expressed early or late in the replication cycle. The 376 major ORFs in chlorella virus PBCV-1 are evenly distributed along the genome, and 85% are separated by less than 200 nucleotides. Likewise, 85% of the 151 putative genes in ASFV are also separated by less than 200 nucleotides (Yanez et al., 1995). The genes in phage T4 are even more tightly packed (Kutter et al., 1994). Consequently, transcription start and stop signals plus the regulatory regions for at least some virus genes are extremely short, or they are located in the coding region of adjacent genes.

(2) Some virus-encoded proteins are smaller than those from free-living organisms and may approach the minimum size required for enzyme activity. Examples include the PBCV-1 encoded 298 amino acid residue ATP-dependent DNA ligase, the 1,061 amino acid residue type II DNA topoisomerase, and the 372 amino acid residue ornithine decarboxylase. Each of these virus-encoded proteins has the expected enzyme activity. ATP-dependent DNA ligases range in size from the 268 amino acid residue enzyme from *Haemophilus influenzae* (Cheng and Shuman, 1997) to the 1,070 amino acid residue enzyme from *Xenopus laevis* (Lepetit et al., 1996). The PBCV-1 enzyme is the second smallest ATP-dependent ligase in the databases. The PBCV-1 encoded type II DNA topoisomerase is about 130 amino acids smaller than the next smallest type II topoisomerase in the databases, which is encoded by virus ASFV (Garcia-Beato et al., 1992). The PBCV-1 encoded ornithine decarboxylase is about 90 amino acids smaller than the next smallest ornithine decarboxylase in the databases. Likewise, the large subunit of ribonucleotide reductase from the baculovirus *Orgyia pseudotsugata* multinucleocapsid nuclear polyhedrosis virus (OpMNpV) is 150 to 200 amino acids smaller than its counterpart from most organisms (Ahrens et al., 1997).

(3) Even though introns were first discovered in adenoviruses (Berget et al., 1977; Chow et al.,

1977), the genes of many large DNA-containing viruses either lack introns, e.g., poxviruses, baculoviruses, iridoviruses, and ASFV, or only have a few short introns, e.g., phycodnaviruses. An absence of introns obviously contributes to the smaller size of virus genes.

To summarize, it is unlikely that studying viruses will reveal useful information about the minimum number and types of genes required to support life. However, the finding that, on average, virus genes can be 10 to 50% smaller than those from bacteria indicate that the minimum genome size required to support life may be smaller than previously thought.

Acknowledgments

I thank Les Lane, Mike Nelson, Myron Brakke, and Mike Graves for their comments on this manuscript and Dan Rock and Gisela Mosig for the information on MsEPV virus and coliphage T4, respectively.

References

Afonso, C.L., E.R. Tulman, Z. Lu, E. Oma, G.F. Kutish, and D.L. Rock. 1998. The genome of *Melanoplus sanguinipes* entomopoxvirus. *J. Virol.* (in press).

Ahrens, C.H., R.L.Q. Russell, C.J. Funk, J.T. Evans, S.H. Harwood, and G.F. Rohrmann. 1997. The sequence of the *Orgyia pseudotsugata* multinucleocapsid nuclear polyhedrosis virus genome. *Virology* **229**:381-399.

Antequera, F., and A. Bird. 1993. CpG Islands. Pp. 169-185 in *DNA Methylation: Molecular Biology and Biological Significance,* P.J. Jost and P.H. Saluz (eds.), Basel, Switzerland: Birkhauser Verlag.

Ayres, M.D., S.C. Howard, J. Kuzio, M. Lopez-Ferber, and R.D. Possee. 1994. The complete DNA sequence of *Autographa californica* nuclear polyhedrosis virus. *Virology* **202**:586-605.

Berget, S.M., C. Moore, and P.A. Sharp. 1977. Spliced segments at the 5'terminus of adenovirus 2 late mRNA. *Proc. Natl. Acad. Sci. USA* **74**:3171-3175.

Cheng, C., and S. Shuman. 1997. Characterization of an ATP-dependent DNA ligase encoded by *Haemophilus influenzae*. *Nucleic Acids Res.* **25**:1369-1374.

Chow, L., R. Gilinas, T. Broker, and R. Roberts. 1977. An amazing sequence arrangement at the 5'ends of adenovirus 2 messenger RNA. *Cell* **12**:1-8.

Dolan, A., F.E. Jamieson, C. Cunningham, B.C. Barnett, and D.J. McGeoch. 1998. The genome sequence of herpes simplex virus type 2. *J. Virol.* **72**:2010-2021.

Doolittle, R.F. 1998. Microbial genomes opened up. *Nature* **392**:339-342.

Fraser, C.M., J.D. Gocayne, O. White, M.D. Adams, R.A. Clayton, R.D. Fleischmann, C.J. Bult, A.R. Kerlavage, G. Sutton, J.M. Kelley, J.L. Fritchman, J.F. Weidman, K.V. Small, M. Sandusky, J. Fuhrmann, D. Nguyen, T.R. Utterback, D.M. Saudek, C.A. Phillips, J.M. Merrick, J.F. Tomb, B.A. Dougherty, K.F. Bott, P.C. Hu, T.S. Lucier, S.N. Peterson, H.O. Smith, C.A. Hutchison, and J.C. Venter. 1995. The minimal gene-complement of *Mycoplasma genitalium*. *Science* **270**:397-403.

Garcia-Beato, R., J.M.P. Freije, C. Lopez-Otin, R. Blasco, E. Vinuela, and M.L. Salas. 1992. A gene homologous to topoisomerase II in African swine fever virus. *Virology* **188**:938-947.

Hutson, M.S., G. Holzwarth, T. Duke, and J.L.Viovy. 1995. Two-dimensional motion of DNA bands during 120° pulsed-field gel electrophoresis. I. Effect of molecular weight. *Biopolymers* **35**:297-306.

Itaya, M. 1995. An estimation of minimal genome size required for life. *FEBS Lett.* **362**:257-260.

Kutish, G.F., Y. Li, Z. Lu, M. Furuta, D.L. Rock, and J.L. Van Etten. 1996. Analysis of 76 kb of the chlorella virus PBCV-1 330-kb genome: Map positions 182 to 258. *Virology* **223**:303-317.

Kutter, E., T. Stidham, B. Guttman, E. Kutter, D. Batts, S. Peterson, T. Djavakhishvili, F. Arisaka, V. Mesyanzhinov, W. Ruger, and G. Mosig. 1994. Genomic map of bacteriophage T4. Pp. 491-519 in *Molecular Biology of Bacteriophage T4*, J.D. Karam (ed). Washington DC: American Society for Microbiology.

Langridge, W.H.R., and D.W. Roberts. 1977. Molecular weight of DNA from four entomopoxviruses determined by electron microscopy. *J. Virol.* **21**:301-308.

Lepetit, D., P. Thiebaud, S. Aoufouchi, C. Prigent, R. Guesne, and N. Theze. 1996. The cloning and characterization of a cDNA encoding *Xenopusa levis* DNA ligase I. *Gene* **172**:273-277.

Li, Y., Z. Lu, D.E. Burbank, G.F. Kutish, D.L. Rock, and J.L. Van Etten. 1995. Analysis of 43 kb of the chlorella virus PBCV-1 330-kb genome: Map position 45 to 88. *Virology* **212**:134-150.

Li, Y., Z. Lu, L. Sun, S. Ropp, G.F. Kutish, D.L. Rock, and J.L. Van Etten. 1997. Analysis of 74 kb of DNA located at the right end of the chlorella virus PBCV-1 330-kb genome. *Virology* **237**:360-377.

Lu, Z., Y. Li, Q. Que, G.F. Kutish, D.L. Rock, and J.L. Van Etten. 1996. Analysis of 94 kb of the chlorella virus PBCV-1 330-kb genome: Map positions 88 to 182. *Virology* **216**:102-123.

Lu, Z., Y. Li, Y. Zhang, G.F. Kutish, D.L. Rock, and J.L. Van Etten. 1995. Analysis of 45 kb of DNA located at the left end of the chlorella virus PBCV-1 genome. *Virology* **206**:339-352.

Mushegian, A.R., and E.V. Koonin. 1996. A minimal gene set for cellular life derived by comparison of complete bacterial genomes. *Proc. Natl. Acad. Sci. USA* **93**:10268-10273.

Reisser, W. (ed). 1992. *Algae and Symbioses*. Bristol, UK: Biopress.

Rohozinski, J., L. Girton, and J.L. Van Etten. 1989. Chlorella viruses contain linear nonpermuted double-stranded DNA genomes with covalently closed hairpin ends. *Virology* **168**:363-369.

Senkevich, T.G., J.J. Bugert, J.R. Sisler, E.V. Koonin, G. Darai, and B. Moss. 1996. Genome sequence of a human tumorigenic poxvirus: Prediction of specific host response evasion genes. *Science* **273**:813-816.

Strasser, P., Y. Zhang, J. Rohozinski, and J.L. Van Etten. 1991. The termini of the chlorella virus PBCV-1 genome are identical 2.2-kbp inverted repeats. *Virology* **180**:763-769.

Tidona, C.A., and G. Darai. 1997. The complete DNA sequence of lymphocystis disease virus. *Virology* **230**:207-216.

Van Etten, J.L., D.E. Burbank, A.M. Schuster, and R.H. Meints. 1985. Lytic viruses infecting a chlorella-like alga. *Virology* **140**:135-143.

Van Etten, J.L., L.C. Lane, and R.H. Meints. 1991. Viruses and viruslike particles of eukaryotic algae. *Microbiol. Rev.* **55**:586-620.

Yamada, T., T. Higashiyama, and T. Fukuda. 1991. Screening of natural waters for viruses which infect chlorella cells. *Appl. Environ. Microbiol.* **57**:3433-3437.

Yanez, R.J., J.M. Rodriguez, M.L. Nogal, L. Yuste, C. Enriquez, J.F. Rodriguez, and E. Vinuela. 1995. Analysis of the complete nucleotide sequence of African swine fever virus. *Virology* **208**:249-278.

SUGGESTIONS FROM OBSERVATIONS ON NANOBACTERIA ISOLATED FROM BLOOD

E. Olavi Kajander, Mikael Björklund, and Neva Çiftçioglu
Department of Biochemistry and Biotechnology
University of Kuopio

ABSTRACT

Nanobacteria are the smallest cell-walled bacteria, only recently discovered in human and cow blood and in commercial cell culture serum. The environment causes drastic changes in their unit size: under unfavorable conditions they form very large multicellular units. Yet, they can release elementary particles, some of which are only 50 nm in size, smaller than many viruses. Although metabolic rates of nanobacteria are very slow, they can produce carbonate apatite on their cell envelope mineralizing rapidly most of the available calcium and phosphate. Nanobacteria belong to, or may be ancestors of, the alpha-2 subgroup of Proteobacteria. They may still partially rely on primordial life-strategies, in which minerals and metal atoms associated with membranes played catalytic and structural roles reducing the number of enzymes and structural proteins needed for life. Simple metabolic pathways and lack of energy-consuming pumps, apparently only compatible with life in very small cells, may support the 10,000-fold slower growth rate (absolute rate of mass gain) of nanobacteria, as compared to the usual bacteria. Simplistic life strategy may also explain the endurability of this life-form in extreme environmental conditions. Nanobacteria may have evolved in environmental sources, e.g., in primordial soups or later as scavengers in hot springs, to take advantage of the steady-state calcium-phosphate and nutrient supply of the mammalian blood. Their elementary particles or units do appear and may function much like viruses, but can support autonomous replication under suitable conditions, e.g., after union of several units, thus opening a new survival strategy for smallest life-forms.

Is There a Relationship Between Minimum Size and Environment?

Nanobacteria and Minimum Size of a Living Cell

Nanobacteria grow under mammalian cell culture conditions. They pass through sterile filters and endure g-irradiation like a virus (1 megarad not effective). Their size is between that of a virus and cell-walled bacteria. They are stained with DNA fluorochromes such as mitochondria. Nanobacteria produce a slimy biomatrix that forms carbonate apatite mineral around them in culture (Kajander et al., 1997; Çiftçioglu et al., 1997, 1998). This bizarre new form of life seems to have adapted to living inside the mammalian body, an ecologically free but hostile niche. The suggested name *Nanobacterium sanguineum* refers to their small size and their habitat, which is blood. Nanobacteria are one of the most distinct organisms ever found in humans. Their poor culturability and long doubling time, and cytotoxicity (Çiftçioglu and Kajander, 1998), can be compared only to some Mycobacteria, such as *M. leprae*. The average diameter of nanobacteria measured with electron microscopic techniques, about 0.2 μm, is smaller than that of large viruses. The smallest units of nanobacteria capable for starting replication in culture, possibly as aggregates of several, have sizes approaching 0.05 μm, based upon filtration and electron microscopic results (Kajander et al., 1997; Çiftçioglu et al., 1997). The theoretical minimum diameter of a cell, based on the size of those macromolecules now considered to be necessary for a living cell, has been calculated to be about 0.14 μm (Himmelreich et al., 1996; Mushegian and Koonin, 1996). Some nanobacterial cells appear smaller than that. Do nanobacteria really exist?

Nanobacteria Do Exist

1. Nanobacteria can be cultured, have a doubling time of about 3 days, and can be passaged apparently forever. Now they have been passaged for over 6 years monthly.

2. They produce biomass at a rate of about 0.0001 times that of *E. coli*.

3. Their biomass contains novel proteins and "tough" polysaccharides.

4. SDS-PAGE of nanobacterial samples shows over 30 protein bands. Amino terminal sequences are available from 6 different proteins One of them is a functional porin protein (unpublished work in collaboration with Dr. James Coulton, McGill University). Porins are a hallmark for gram-negative bacteria located in their outer membrane and make trafficking through it possible for relatively small molecules. Porins seem to be located in the mineral layers in nanobacteria. Muramic acid, a major component of peptidoglycan, has also been detected. So, nanobacterial cell walls do have typical gram-negative components, although their ultrastructure is unique and varies during their growth phases.

5. Nanobacteria contain modified nucleic acids detectable specifically with stainings and spectroscopy, and their components can be detected with mass spectroscopy (Kajander et al., 1997).

6. Nanobacterial growth can be prevented with small concentrations of tetracycline antibiotics, or with high concentrations of aminoglycoside antibiotics. Both stop bacterial protein synthesis at the ribosomal level.

7. Nanobacterial growth can be prevented with small concentrations of cytosine arabinoside or fluoro-uracil, both of which are antimetabolites preventing nucleic acid synthesis in all types of cells.

8. Nanobacteria can be detected with metabolic labeling using methionine or uridine.

9. Nanobacteria have unique strategies for social behavior and for multiplication, including communities, budding, and fragmentation.

Nanobacterial Mineral Is Biogenic

All carbonate apatite in the human body is biogenic. Nanobacterial mineral formation is a specific biogenic process, for these reasons:

1. Mineral grows directly on the nanobacteria, forming parts of the cell envelope. Without nanobacteria there is no mineralization in the medium. Mineral growth is dependent on a biomatrix made by the nanobacteria (Kajander and Çiftçioglu, 1998).

2. Mineral layer is under active remodeling of its size and shape, and it is budding.

3. No significant mineralization takes place if nanobacteria are killed with γ-irradiation.

4. Mineralization is an active process that does not imply supersaturation. It brings phosphate levels to zero in the culture medium (Kajander et al., 1998).

5. Mineral grows as layers in a biomatrix, comparable to that in pearls.

6. Mineral crystallization is under biocontrol with serum factors, much as bone is.

Nanobacteria Are Distinct Bacteria and Not "Contaminants" of Biological Samples

We have found nanobacteria belonging to, or being an ancestor of, a group of bacteria, the alpha-2 subgroup of Proteobacteria, that contain both environmental bacteria and bacteria inhabiting mammalian blood and tissues. The nearest relatives are Phyllobacteria found in soil and causing tropical plant diseases. These bacteria do not produce apatite and differ much from nanobacteria (Table 1).

Table 1 Nanobacteria Compared to Phyllobacteria, Their Closest Relatives in 16S rRNA Gene Comparison

Nanobacteria	Phyllobacteria
Culturable only in cell culture medium	Culturable in most bacterial media
Thermophile, gamma-irradiation resistant	Maximum growth temperature 32º C, gamma sensitive?
Present in blood, very slow grower	Present in soil and plants, fast grower
Mineralizing, ultrastructure is unique	No minerals, ultrastructure is gram negative
No polyamines, but cadaverine-like compound	Normal polyamines present
Modified nucleic acid bases present	Normal nucleic acid components
Specific protein pattern, sequences, epitopes	Specific protein pattern, sequences, epitopes
Porin protein only weakly cross-reactive	Porin protein only weakly cross-reactive
Polymerase chain reaction (PCR) needs special protocol	PCR works with standard protocols

Nanobacteria and the Other Small Bacterial Forms

Bacteria do exist in sedimentary rocks. Much of this bacterial metabolism and function is unlike that of previously known organisms, and is related to the slow mineralization of inorganic and organic compounds available. From such biota, particles resembling our tiniest nanobacteria were discovered by Dr. Folk, who named them as "nannobacteria" (Folk, 1993). They may contribute to the formation of carbonate minerals and remain uncharacterized. Ultramicrobacteria, passing through sterile filters, have been found in soil and natural water sources. They are difficult to culture and their nature is largely unknown (Roszak and Colwell, 1987), as is their possible connection to nanobacteria. Normal bacteria may acquire a dormant state and do not even multiply on subsequent culture (Roszak and Colwell, 1987). The size of such starved cells can be only a fraction of the size obtained when multiplication is reached again. Nanobacteria are not in a dormant state.

Cell-wall-deficient bacteria, L-forms, show small and large forms. Conventional culture methods do not support the growth of L-form microbes. L-forms can pass through sterile filters but can be easily lysed and their nucleic acids and proteins extracted (Darwish et al., 1987). *Mycoplasma, Chlamydia*, and *Rickettsia* are the smallest "classically known" bacteria, and they can be cultured in cell culture conditions with mammalian cells. Only mycoplasma can grow autonomously. All can pass through sterile filters: filtering through 0.2 µm pore-size results in over 100-fold reduction in their numbers, whereas with nanobacteria the reduction is typically less than 10-fold (Kajander et al., 1997), and bacterial L-forms are reduced by 10^6-fold (Darwish et al., 1987).

"Pseudoorganisms" forming "pseudocolonies" have been detected in mycoplasma culture media. These were regarded as non-living artifacts, e.g., calcified fatty acids, owing to resistance to disinfectants and unsuccessful attempts at DNA detection (Hijmans et al., 1969). Some of their properties were similar to those of nanobacteria: presence in serum, difficulties in fixation or in disruption, inability to stain with common dyes, resistance to antibiotics and disinfectants, and high calcium-phosphate content. Buchanan (1982) found similar "pseudocolonies" in several horse sera but considered them as atypical bacterial L-forms.

Size is considered to be typical for a certain bacterial species. The alternative is that size, shape, and morphology change according to the environmental and social status of the organism. Examples of such organisms are known. *Myxococcus xanthus* has a life cycle, carefully controlled by cell density and nutrient levels, and consisting of tiny forms, actively moving large forms, and huge social formations producing mushroom-like fruiting bodies. Nanobacteria do show several growth forms, sizes, and social formations depending on culture conditions. Fastly growing mycoplasma "forget" cell division, forming very long multicellular forms. Thus, bacterial size is dependent on growth phase. Small size is not directly linked to the genomic size: *Myxococcus xanthus* genome size 9.4 Mb (Chen et al., 1990) is among the largest, whereas mycoplasmas have the smallest genome sizes, 0.58-1.6 Mb (Barlev and Borchsenius, 1991). *Chlamydia* and *Rickettsia* have genomes of 1 Mb. Nanobacterial genome size is unknown, but quantitative Hoechst staining suggests it may be smaller than that of mycoplasmas.

Is There a Continuum of Size and Complexity That Links Conventional Bacteria to Viruses?

Nanobacteria, Mycoplasma, Chlamydia, and *Rickettsia* are structurally only a little more complex than large viruses. They all use environmental supplies appropriately to minimize the need for their own synthetic pathways. Nanobacterial cultures do indicate virus-sized elementary particles and large nanobacteria acting like mother cells in a life cycle involving nonreplicative and replicative forms. This is analogous to modern gene technology: viruses, helper viruses, and competent bacteria are used to replicate new viral particles.

Simplistic Strategies by Nanobacteria

Nanobacterial function is simple: be ready for nutrients when they come, replicate, make protective mineral to "hibernate," and wait for a new cycle of nutrients. The main features are these:

1. Nanobacteria use ready amino acids from medium/environment.
2. They use large amounts of Gln, Asn, and Arg from medium for structural components, or energy production or mineralization process (amino groups could bind phosphate).
3. They use ready fatty acids from their medium. When fatty acids are scarce, they are "saved" by replacing membrane lipids partly with apatite.
4. They react to stress by becoming social and forming communities. Communities may help to overcome mutations, etc. They can "hibernate" for extensive periods waiting for suitable conditions permitting growth.
5. Because of their small size, nutrients can be obtained by diffusion and brownian movements.
6. Nanobacteria may have low internal pressure. Normal bacteria concentrate metabolites inside them so that their internal pressures can be 3-5 bars. Such a system provides fast metabolism, but consumes energy and requires complex pumps and their controls. In unfavorable conditions cell death can result from inability to keep up the ion gradients. Nanobacteria may lack these systems. That might explain partially their high resistance to near-boiling temperatures (Björklund et al., 1998) known to explode bacteria mainly owing to an imbalance in intracellular ions. Their endurance is similar to that of some viruses.
7. Nanobacteria may form and shed units resembling viruses that could spread even via tiny pores or cracks, e.g., in rocks.

The survival strategy of nanobacteria indicates that small is efficient in these ways: minimize synthetic systems, energy consumption, pumps; scavenge nutrients when they are available; endure deadly attacks but eat up nutrients from dead bystanders; and have a strategy for surviving in very hostile places that kill normal bacteria (hot springs) or places providing all nutrients (primordial soup, blood).

What Is the Phylogenetic Distribution of Very Small Bacteria?

The most powerful comparison can now be based upon genomic sequences of organisms. Mycoplasmas are among the smallest bacteria, with a diameter of about 0.2-0.5 µm, and their genomic size is the smallest so far known. The *M. genitalium* genome is 0.58 Mb compared to 4.6 Mb for *E. coli*. The small genome seems to be an indicator of life strategy, the parasitic life style. Such organisms do not need to manufacture all their building blocks themselves. Could this apply for environmental simplistics? What type of metabolic simplifications could be possible?

Polyamines and Life Strategy

Polyamines are now considered essential for cell proliferation. Bacteria contain putrescine and spermidine, but may contain some 30 other di- and polyamines. Their patterns have been used as a phylogenetic tool (Hamana and Matsuzaki, 1992). What can be learned on the enzymes of polyamine synthesis from the genomic sequences? Genes for enzymes producing putrescine and spermidine are absent in *M. genitalium*, *Borrelia burgdorferi,* and *Treponema pallidum*. *Haemophilus influenza*e can produce putrescine, and *Helicobacter pylori*, *Mycobacterium tuberculosis*, and *E. coli* can produce both putrescine and spermidine. Some Archaea, Methannococcus (*M. jannaschii*) and Halococcus, lack synthesis of polyamines and lack them in direct analysis (Hamana and Matsuzaki, 1992). Nanobacteria do not have putrescine or spermidine, but contain a compound having similar mobility with cadaverine in high pressure liquid chromatography. Cadaverine, a special polyamine used by several eubacteria as a covalently linked component in peptidoglycan, absence of normal eubacterial polyamines, and lack of putrescine/spermidine transporter genes make nanobacteria unique. The parasitic bacteria acquire their polyamines from their hosts, and can thus afford losing the synthetic enzymes of importance to their freely living relatives. The environment provides compensation for the loss. What is the smallest genetic size for life? Obviously it depends on the generosity of the environment and the life strategy.

Smaller Is More Practical

Organisms must have been very small in primordial soups! And slow growers. Large cells would have to have complex systems including active transporters and moving apparatus. Small cells can rely on diffusion and Brownian movements for obtaining nutrients. Very slow metabolic rates would allow for use of minimal numbers of enzymes, since many of the reactions could be uncatalyzed, or catalyzed by metals and minerals or be contributed by nonspecificity of the existing enzymes. Such a system may well do the observed 10,000-fold slower biomass production than that of common bacteria. Nanobacteria have apparently small genomes. Hoescht 33258 staining indicated that nanobacteria should have DNA amounts between that of mycoplasmas and mitochondria. Can bacteria have novel nucleic acids contributing to smallness? One potential example could be use of single stranded nucleic acid genome, maybe resembling the multi-copy single stranded DNA found in bacteria.

Further simplification would be obtained by omitting the need for a closed compartment needed to keep homeostatic conditions intracellularly. We are suggesting an elementary system of tiny units performing special tasks. Only when united and surrounded by membrane, closing the compartment, would they resemble present forms of bacteria.

Mitochondria in *Saccharomyces cerevisiae* have 35 genes, and about 290 more are in the nuclear genome (Hodges et al., 1998). So mitochondria are operating probably with a smaller number of genes—but with a full operational capability—than any modern bacteria. Mitochondria would fall into the alpha-2 subgroup of Proteobacteria, if classified as bacteria, and thus be near-relatives of nanobacteria. They may have lost many genes in the process of domestication as a eukaryotic cell organelle. This also points out that metabolic collaboration between various bacteria, or bacteria and other organisms, can significantly reduce necessary genomic sizes. This is understood from the fact that none of the bacteria with genomic sizes 1.6 Mb or smaller can synthesize polyamines necessary for their growth. The suggested minimum number of genes, 256 genes (Mushegian and Koonin, 1996), may be still too high a number for the simplest genome for the reasons discussed above. Another conclusion is that it is possible to evolute into miniature life-forms from several bacteria groups, since the smallest organisms fall into several classes. The main factor for thriving is the environment and stability of its conditions: primordial soup may have provided nutrients for supporting organisms with many fewer genes than are necessary to survive in present-day environments. Why do we think that nanobacteria may serve as a model for primordial life? Because they may well be just that! The modern-day primordial soup is blood.

References

Barlev N.A., and S.N. Borchsenius. 1991. Continuous distribution of Mycoplasma genome sizes. *Biomed. Sci.* **2**:641-645.

Björklund M., N. Çiftçioglu, and E.O. Kajander. 1998. Extraordinary survival of nanobacteria under extreme conditions. *Proceedings of SPIE* **3441**:123-129.

Buchanan A.M. 1982. Atypical colony-like structures developing in control media and clinical L-form cultures containing serum. *Vet. Microbiol.* **7**:1-18.

Chen H., I.M. Keseler, and L.J. Shimkets. 1990. Genome size of *Myxococcus xanthus* determined by pulsed-field gel electrophoresis. *J. Bacteriol.* **172**:4206-4213.

Çiftçioglu N., A. Pelttari, and E.O. Kajander. 1997. Extraordinary growth phases of nanobacteria isolated from mammalian blood. *Proceedings of SPIE* **3111**:429-435.

Çiftçioglu N., and E.O. Kajander. 1998. Interaction of nanobacteria with cultured mammalian cells. *Pathophysiology* **4**:259-270.

Çiftçioglu N., M. Björklund, and E.O. Kajander. 1998. Stone formation and calcification by nanobacteria in human body. *Proceedings of SPIE* **3441**:105-111.

Darwish R.Z., W.C. Watson, M.R. Belsheim, and P.M. Hill. 1987. Filterability of L-forms. *J. Lab. Clin. Med.* **109**:211-216.

Folk R.L. 1993. SEM imaging of bacteria and nannobacteria in carbonate sediments and rocks. *J. Sediment. Petrol.* **63**:990-999.

Hamana K., and S. Matsuzaki. 1992. Polyamines as a chemotaxonomic marker in bacterial systematics. *Crit. Rev. Microbiol.* **18**:261-283.

Hijmans W., C.P.A. van Boven, and H.A.L. Clasener. 1969. Fundamental biology of the L-phase of bacteria. Pp.118-121 in *The Mycoplasmatales and L-phase of Bacteria*, L. Hayflick (ed.). New York: Appleton-Century-Crofts.

Himmelreich R., H. Hilbert, H. Plagens, E. Pirkl, B.C. Li, and R. Herrmann. 1996. Complete sequence analysis of the genome of the bacterium *Mycoplasma pneumoniae*. *Nucleic Acids Res.* **24**:4420-4449.

Hodges P.E., W.E. Payne, and J.I. Garrels. 1998. Yeast Protein Database (YPD): a database for the complete proteome of *Saccharomyces cerevisiae*. *Nucleic Acids Res.* **26**:68-72.

Kajander E.O., and N. Çiftçioglu. 1998. Nanobacteria: An alternative mechanism for pathogenic intra- and extracellular calcification and stone formation. *Proc. Natl. Acad. Sci. USA* **95**:8274-8279.

Kajander E.O., I. Kuronen, K. Åkerman, A. Pelttari, and N. Çiftçioglu. 1997. Nanobacteria from blood, the smallest culturable, autonomously replicating agent on Earth. *Proceedings of SPIE* **3111**:420-428.

Kajander E.O., M. Björklund, and N. Çiftçioglu. 1998. Mineralization by nanobacteria. *Proceedings of SPIE* **3441**:86-94.

Mushegian A.R., and E.V. Koonin. 1996. A minimal gene set for cellular life derived by comparison of complete bacterial genomes. *Proc. Natl. Acad. Sci. USA* **93**:10268-10273.

Roszak D.B., and R.R. Colwell. 1987. Survival strategies of bacteria in the natural environment. *Microbiol. Rev.* **51**:365-379.

PROPERTIES OF SMALL FREE-LIVING AQUATIC BACTERIA

D.K. Button[1,2] and Betsy Robertson[1]
[1]Institute of Marine Science and [2]Department of Chemistry and Biochemistry
University of Alaska at Fairbanks

Abstract

The smallest genome size for free-living cell-wall defined bacteria is ~ 1 Mb (1.1 fg DNA)/cell. Lower limits of genome size appear to be affected by forces favoring nutritional complexity in dilute aquatic systems where small size gives a favorable surface to volume ratio for nutrient collection. Evolutionary forces, according to thermodynamic principles, tend away from extremely small genomes, and these are not known among the bacteria. Space for the DNA to undergo required conformational changes probably affects minimum organism size. The minimum cell mass for cultured bacteria appears to be 25 fg dry weight for cultured bacteria, and about 10 fg for those in aquatic systems. Additional space for DNA replication in some small cells is provided for by a dilute cytoplasm.

Introduction

Heterotrophic bacteria number from 10^4 to 10^6 ml throughout most aquatic systems. As a consequence of persistence throughout evolutionary history, and with rapid rates of reproduction and without gene duplication to minimize mutation, bacterial sizes have undergone unprecedented periods of adaptation with presumably little change in morphology. Sizes of the smaller aquatic bacteria have been based on dimensions of stained cells according to epifluorescence microscopy, electron microscopy of prepared specimens (Loferer-Krössbacher et al., 1998), and most recently from the intensity of scattered light (Robertson and Button, in progress) by flow cytometry. These transthreptic (across-surface feeding) chemoheterotrophic aquatic bacteria are small and generally difficult to grow. Recent attempts to culture typical aquatic bacteria from single cells in unamended seawater (Button et al., 1993), together with kinetic (Button, 1998) and flow cytometric analysis (Button and Robertson, 1993) of the resulting populations have improved understanding of these small organisms. Here we discuss those results from the perspective of minimal attainable size.

Methods

The light scatter attending single cells can be analyzed by flow cytometry, separating the bacteria from one another hydrodynamically and from debris according to the intensity of DNA-specific stains. Because the DNA content is large, from 1.5 to 8 fg/cell for most aquatic bacteria, there is little interference from debris and other organisms, and observed signals are thought to emanate almost entirely from the bacteria. The scatter signal depends on the dipole content of the particle and therefore is a reflection of dry mass (Robertson et al., 1998). There is a shape dependency, but this becomes negligible for the smaller organisms. Samples require formaldehyde treatment for stain penetration and preservation which adds to the dry mass. Some variation occurs in the amount of formaldehyde absorbed among species, but by using aquatic bacteria thought to be typical as standards, this error is minimized. The increase was 15% for *Escherichia coli* and 35% for the more dilute extinction culture isolate *Cycloclasticus oligotrophus,* which has a smaller bouyant density as well. Dry weights of the latter can be calculated from the radoactivity acquired from labeled substrates and used to calibrate the light scatter curve from flow cytometry. The anticipated curve is obtained from light scatter theory, and the cell volumes of *E. coli* are large enough for accurate measurement by electronic impedance. Values for the two methods agree. Experimental data agree with the curve calculated from light scatter theory and instrument geometry, and since the theory can be extended to include the smallest of bacteria, flow cytometry can be used to measure their dry mass. They also confirm significant shrinkage in cells measured by electron microscopy.

To obtain representative marine bacteria and estimate their viability, seawater was diluted to a single reproducing cell and the resulting culture statistically (Quang et al., 1998) and physiologically (Schut et al., 1993; Button et al., 1998) characterized in a procedure called dilution culture.

The Smallest Bacteria

Signatures according to flow cytometric profiles vary according to depth and location, but the differences are not great. Taking a typical example, such as at 500 m from the Gulf of Alaska, values for the mean, median, and mode for dry mass are 0.0171, 0.0113, and 0.0123 pg/cell. For volume they are 0.100, 0.0667, and 0.0722 μm^3/cell; 95% of the population is >0.028 μm^3/cell in volume assuming 20% dry weight. The distinction between very small bacteria and instrument noise is uncertain, but filtered controls show little signal. Assuming this smallest component is either non-bacterial or at the most 2% of the biovolume, the smallest known bacteria are of the order of 0.4 μm in diameter with the possible exception of *Nanobacterium sanguineum* discussed in the proceedings from this workshop.

Bacterial Size and Genome Size

According to flow cytometry data, genome sizes of cultivated aquatic bacteria range upward from the lowest known for a conventional isolate, 1.6 Mb for *Pseudomonas* [*Brevundimonas*] *diminuta*. Extinction culture isolates are intermediate at 1.6 for *Sphingomonas* sp. RB 2256 (Schut et al., 1993) and 2.2 Mb for *Cycloclasticus arcticus*. All values are smaller than for *Escherichia coli* with a genome size of 4.6 Mb. Mycoplasma have the smallest genomes of the reproducing organisms at 0.6 Mb and are cell-wall free (Krawiec and Riley, 1990). Such organisms would appear near the low fluorescence or "dim" fraction of marine organisms.

Most marine bacteria are 10 to 50 fg/cell dry mass with DNA ranging from 1.2 to 6 fg/cell in rough proportion to size. The DNA level in cultivatable marine bacteria is high: 11% of the dry mass for *C. oligotrophus* compared with 1.6% for *E. coli*. The smaller half of the marine population has a dry mass of 10 to 20 fg/cell while DNA ranges from 1.5 to 2 fg/cell. The DNA content of the cells in this group ranges from 5% of the dry mass at the large end of the size distribution to 15% for those at the small end.

A distinguishing characteristic of the small-genome *Brevundimonas diminuta* is its ability to grow on only a few sugars and amino acids (Segers et al., 1994). Alternatively, *E. coli* can use many substrates, and it has 285 (Joel et al., 1994) or more (considering unknown sequences) different transport systems to accumulate them. *C. oligotrophus* uses primarily hydrocarbons that number many in type and the complexities of their metabolism are unknown. We therefore conclude that genome size is related to nutritional complexity, that the minimum genome size in aquatic systems is about 1 fg/cell, and that cell size is limited by genome size when the latter is of the order of 15% of the organism's dry weight.

Size Distributions with Predation

Bacterivores are thought to prefer larger bacteria (Sherr et al., 1992); however, grazing is reported to invigorate populations, leading to faster rates of growth and concomitant increases in bacterial size (Han and Höfle, 1998). Experiments that compared bacterial sizes in seawater passed through 0.45 mm filters to remove the bacterivores, followed by incubation in dialysis bags, resulted in populations having the same size distribution by flow cytometry and failed to show the anticipated grazing effect (Button and Robertson, unpublished). Moreover, while significant changes in bacterial size are often reported with depth, cell sizes by flow cytometry are relatively constant from the surface to several thousand meters where minimal grazing is expected.

Cytoarchitectural Reflections in Kinetic Constants

Specific affinity theory relates nutrient sequestering ability to cytoarchitecture. The rate constant for nutrient collection by a cell, the specific affinity, a_S, depends on the number of permease molecules N on the cell surface,

$$a_s = \frac{Nc}{1 + Sc\tau}, \qquad (1)$$

the concentration of substrate S, and the residence time τ. Tau is the mean time the substrate spends with the pathway enzymes between uptake and utilization including steps that can retard movement of molecules through the pathway following successful substrate collision with the permease, and c is constant dependent on substrate and organism properties (Button, 1998). Nutrient sequestering power depends on N and indirectly on surface to volume ratio, and it increases with reciprocal cell size. The affinity constant K_A, where $S c \tau$ is unity and the specific affinity is half its base value a^o_S, decreases with τ. The result is a reciprocal relationship between K_A and a^o_S. Organisms adapted to dilute environments have small affinity constants and large specific affinities. This is thought to reflect large numbers of permease molecules compared with cytoplasmic enzymes, because small numbers of cytoplasmic enzymes increase τ. The small amounts of cytoplasmic enzymes required conserve space, resulting in a dilute cytoplasm that minimizes endogenous requirements and favors small cell size.

Permease Diversity

The collisional-limit theory has the rate of nutrient molecule collection independent of the permease content of the cells for molecules that diffuse to within a short distance from the cell surface (Berg and Purcell, 1977; Abbott and Nelsestuen, 1988) unless permease molecules are few. If a nutrient molecule is likely to be collected by a permease distributed across the cell surface at only moderate concentrations, the cell may better devote the space, material, and energy to collection of a different nutrient (Button, 1994), so long as the requirement for increased numbers of catabolic pathways does not outweigh the advantage of multiple substrate use. The competitive advantage of multiple substrate use can be seen in Equation 2. In the cases examined (*Marinobacter arcticus* and *Sphingomonas* sp.), specific affinities a^o_S for amino acids as a group are much larger than those for single amino acids alone. Since growth rate m is given by

$$\mu = \sum_{i=1}^{n} a^o{}_{Si} S_i Y_i, \qquad (2)$$

where Y_i is the cell yield for each substrate S_i used, multiple simultaneous use is of clear advantage. However, there is the cost of maintaining and enclosing additional genes. The cost of additional permeases in cytoplasmic enzymes is less certain since less degradative enzymes may be required, and the excess requirement could be offset by the need for fewer anabolic enzymes. The trend toward smaller dry mass with more limited substrate can be seen from a few examples. *Escherichia coli* uses very many substrates, accumulates them with hundreds of permeases as mentioned (Joel et al., 1994), and has a cell volume of 2.3 µm³. *Marinobacter arcticus* has a cell size of 0.19 µm³ and uses a large number of substrates. *Cycloclasticus oligotrophus* uses several substrates and has a cell size of 0.17 µm³. And *Brevundimonas diminuta* also has a cell size of about 0.17 µm³ and uses only very few substrates (Segers et al., 1994). So, existence in a dilute, multiple substrate environment favors cells that are not as small as they otherwise might be, both because additional space is required for the large genome associated with the additional permeases, and so that a maximum surface to volume ratio can be achieved.

Thermodynamic Driving Forces

In aquatic systems there is a steady input of dissolved organics from phytoplankton that is steadily removed by bacteria. Decomposition of the organics provides minerals to the phytoplankton resulting in a cycle. Concentrations of the organics approach a minimum value set by thermal and apoptostic decomposition processes in the organisms that set maximum lifetimes of the bacteria. One of the forces governing the way in which organisms evolve to meet a complex nutritional system is by generating complexity (Wicken, 1980; Prigogine and Stengers, 1984). It is the result of a tendency of evolving self-perpetuating systems toward minimum rates of entropy production. One might therefore not be surprised to find some very small organisms with the required small genomes in mature aquatic systems. In fact the nutrient-collection ability of a microorganism with the largest specific affinity known is only a few percent of the maximal theoretical value (Button et al., 1998), so extremely small forms are selected against. Sheltered systems could contain smaller organisms provided they are bathed in a mixture of suitably rich diversity.

Diluteness

Measurements exist, using the current technology, for determining the dry-matter content of two marine bacterial species (Robertson et al., 1998), and both values were small. At about 20% dry weight, these organisms contained about a third less dry material than *E. coli*. A dilute cytoplasm is advantageous in that less material is provided to bacterivores in pelagic environments that must seek their prey one organism at a time. Over long periods one might expect predators of free-living bacteria to prefer larger cells with a high percentage of dry material, and perhaps attached into short chains, because well-dispersed meals of 10^{-14} g are not very energizing and equilibrium concentrations of nutrients are reduced. Conversely, the swimming distance between prey is increased. Whatever the effect of selective predation, a larger cytoplasmic cavity becomes available for genome arrangements with increasing water content of the organisms, a critical factor when DNA content approaches 15% dry weight. Costs include a requirement for material and energy to construct the additional cell wall material required to enclose the organisms.

Regulation

One might presume, given the dilute and rather constant chemical composition of seawater, that the metabolic pathways of marine bacteria would not overload and the organisms could rely on kinetic control through the various permease types to supply nutrients at the required level and ratio. For known isolates this is not the case, and induction of numerous membrane and cytoplasmic proteins occurs (Button et al., 1998; Button et al., in preparation). Therefore it appears that space and material are devoted to regulation for many oligobacteria. However, detailed properties of the smallest bacterial fraction of aquatic systems are unknown.

Viability

It can be questioned whether the smallest bacteria measured by flow cytometry are reproducing cells. Dim organisms of low DAPI-DNA fluorescence appear and can be formed from active cultures by starvation (Robertson et al., 1998). Viabilities obtained by dilution to extinction are about 3% in summer (Button et al., 1993), and are possibly higher in the fall. Even with new diffusion culture techniques whose chemical changes are minimized, viabilities are only about 1% in productive Gulf of Alaska waters in the spring (Button et al., unpublished). On the other hand, about half are autoradiography-positive for single amino acids, and small cells sorted by flow cytometry, following incubation with radiolabeled amino acids alone, were as radioactive as the larger fraction, indicating that the small cells are not slowly losing biomass and activity. The lowest-DNA (dim) cells persist in surface waters where grazing would be expected to remove nonreproducing forms. Furthermore, enclosure in sample bottles or increasing the temperature can be vastly stimulatory. Evidence suggests that a large portion of the particles that appear as bacteria, excluding all those above virus in size, are alive.

References

Abbott A.J., Nelsestuen G.L. (1988), The collisional limit: an important consideration for membrane-associated enzymes. *FASEB Monogr* **2**:2858-2866.
Berg H.C., Purcell E.M. (1977), Physics of chemoreception. *Biophys J* **20**:193-219.
Button D.K. (1994), The physical base of marine bacterial ecology. *Microb Ecol* **28**:273-285.
Button D.K. (1998), Nutrient uptake by microorganisms according to kinetic parameters from theory as related to cyto-architecture. *Microbiol Mol Biol Rev* **62**(3):636-645.
Button D.K., Robertson B.R. (1993), Use of high-resolution flow cytometry to determine the activity and distribution of aquatic bacteria. *Handbook of Methods in Aquatic Microbial Ecology,* Kemp P.F., Sherr B.F., Sherr E.B., Cole J.J. (eds). Ann Arbor, Michigan: Lewis.
Button D.K., Robertson B.R., Schmidt T., Lepp P. (1998), A small, dilute-cytoplasm, high-affinity, novel bacterium isolated by extinction culture that has kinetic constants compatible with growth at measured concentrations of dissolved nutrients in seawater. *Appl Environ Microbiol* **64**:3900-3909.
Button D.K., Schut F., Quang P., Martin R.M., Robertson B. (1993), Viability and isolation of typical marine oligobacteria by dilution culture: Theory, procedures and initial results. *Appl Environ Microbiol* **59**:881-891.
Han M., Höfle M.G. (1998), Grazing pressure by a bacterivorous flagellate reverses the relative abundance of *Comamonas acidovorans* PX54 and *Vibrio* strain CB5 in chemostat cocultures. *Appl Environ Microbiol* **64**:1910-1918.
Joel J.J.Y., Cui X., Reizer J., Saier M.H.J. (1994), Regulation of the glucose:H^+ symporter by metabolite-activated ATP-dependent phosphorylation of HPr in *Lactobacillus brevis*. *J Bacteriol* **176**:3484-3492.
Krawiec S., Riley M. (1990), Organization of the bacterial chromosome. *Microbiol Rev* **54**:502-539.
Loferer-Krössbacher M., Klima J., Psenner R. (1998), Determination of bacterial cell dry mass by transmission electron microscopy and densitometric image analysis. *Appl Environ Microbiol* **64**:688-694.
Prigogine I., Stengers I. (1984), *Order Out of Chaos*. Toronto: Bantam.

Quang P., Button D.K., Robertson B.R. (1998), Use of species distribution data in the determination of bacterial viability by extinction culture of aquatic bacteria. *J Microbiol Methods* **33**:203-210.

Robertson B.R., Button D.K. (in progress), Bacterial biomass from measurements of forward light scatter intensity by flow cytometry. *Current Protocols in Cytometry,* Robinson P. (ed.). New York: John Wiley & Sons.

Robertson B.R., Button D.K., Koch A.L. (1998), Determination of the biomasses of small bacteria at low concentration in a mixture of species with forward light scatter measurements by flow cytometry. *Appl Environ Microbiol* **64**:3900-3909.

Schut F., DeVries E., Gottschal J.C., Robertson B.R., Harder W., Prins R.A., Button D.K. (1993), Isolation of typical marine bacteria by dilution culture: Growth, maintenance, and characteristics of isolates under laboratory conditions. *Appl Environ Microbiol* **59**:2150-2160.

Segers P., Vancanneyt M., Pot B., Toruck U., Hoste B., Dewettinck D., Falsen E., Kersters K., DeVos P. (1994), Classification of *Pseudomonas diminuta* Leifson and Hugh 1954 and *Pseudomonas vesicularis* Büsing, Döll, and Freytag 1953 in *Brevundimonas* gen. nov. as *Brevundimonas diminuta* comb. nov. and *Brevundimonas vesicularis* comb. nov., respectively. *Int J Syst Bacteriol* **44**:499-510.

Sherr B.F., Sherr E.B., McDaniel J. (1992), Effect of protistan grazing on the frequency of dividing cells in bacterioplankton assemblages. *Appl Environ Microbiol* **58**:2381-2385.

Wicken J.F. (1980), A thermodynamic theory of evolution. *J Theor Biol* **87**:9-23.

BACTERIA, THEIR SMALLEST REPRESENTATIVES AND SUBCELLULAR STRUCTURES, AND THE PURPORTED PRECAMBRIAN FOSSIL "METALLOGENIUM"

James T. Staley
Department of Microbiology
University of Washington at Seattle

Abstract

The smallest members of the domain Bacteria known to date are found in the following phylogenetic groups: Proteobacteria, Chlamydia, Gram-positive bacteria, Spirochetes, and Verrucomicrobia. The Spirochetes contain very thin bacteria with some species having cell diameters of about 0.1 to 0.15 μm that are at least 5 to 6 μm in length. Apart from this group, the author is not aware that any of other phylogenetic groups produce cells or buds that are less than 0.2 to 0.25 μm in diameter. Likewise, buds, baeocytes, resting, and dispersal stages such as spores and cysts are not known to be less than 0.25 μm in diameter.

Subcellular bacterial structures, such as fimbriae, gas vesicles, prosthecae, and stalks may be as small as 5 to 10 nm in diameter. Some of these are released from cells into environments and may become fossilized. However, the author is not aware that any such structures have ever been reported as fossils even though the remnants of some structures, such as the heavily encrusted stalk of *Gallionella*, would appear to be excellent candidates for this. The search for and verification of fossils of small, single-celled microorganisms and subcellular microbial structures is warranted.

"Metallogenium" is the name given to a structure of microbial size found in the hypolimnion of lakes. This heavily salified rosette structure has been regarded as a bacterium by some, but current evidence suggests that it is non-cellular.

Introduction

Prokaryotic cells show a tremendous range in size. The largest known bacterium is *Thiomargarita*, the denitrifying sulfur-oxidizer found off the west coast of southern Africa; its cells are over 500 μm in diameter. However, such large cell sizes are a rarity in the prokaryotic world.

Certain physical constraints dictate the minimum size of an organism. All cells have a cell membrane, cytoplasm, ribosomes, and nuclear material. Cell membranes are about 8 to 10 nm thick, and sufficient DNA, ribosomes, and enzymes are needed for cells to metabolize and reproduce.

The cell size of many bacterial species is variable, being influenced by growth conditions. Actively growing cells of bacteria are typically larger than senescent cells, and starving cells may be very small indeed. In fact, it is possible that starving cells may turn over so much of their cell matter that they are no longer able to reproduce, and therefore persist in the environment as nanocarcasses less than 0.2 μm in size. From a macromolecular perspective, these organisms would be expected to be depleted in RNA and protein, but rich in DNA. The finding of high concentrations of DNA in particulate materials from natural oligotrophic environments (e.g., Holm-Hansen et al., 1968) is a likely indication that many of the bacteria in such environments are either growing at very low rates or not growing at all.

Also, the effects of physical parameters may be very important in determining cell sizes. Factors such as gravity, pressure, pH, and temperature may influence cell sizes during evolution and selection.

Of course, if the question is, what is the smallest size of a living entity, then bacteria may not be our best example. It is possible that the smallest living entities are precellular. Thus, if life-forms on other planets are different from those on Earth, bacteria may not be the ideal model for comparison.

Selective Advantages of Small Size

Fossil and geochemical records indicate that microorganisms have existed successfully on Earth for more than 3.5 Ga. Indeed, they have persisted despite the evolution of morphologically complex macroorganisms. This observation suggests that there are certain selective evolutionary advantages of small size. Conceivably, small organism sizes could be selected because of (a) the exploitation of niches found in microenvironments, (b) parasitism, (c) oligotrophy, and (d) production of small reproductive cells and spores. Each of these potential selective advantages is discussed briefly below.

Exploitation of Niches Found in Microenvironments

Abundant evidence indicates that microorganisms flourish in microenvironments that are too small to be exploited by macroorganisms. For example, narrow vertical gradients of sulfide and light found in intertidal marine sediments have selected for microbial mat communities structured in millimeter-thick strata. Likewise, anaerobic sediment gradients in which alternate electron acceptors exist are dominated by various bacterial groups involved in fermentations and anaerobic respirations. The microbial loop, which consists of various microbial groups that ingest and degrade microorganisms and small detritus particles, is another example of a microenvironment. However, although these microenvironments are small, they are not of nanometer size, and there are no specific examples of microorganisms less than 0.1 to 0.25 μm in diameter that are known to occupy such a habitat.

Parasitism

Parasites rely on host organisms for materials and in some cases even energy generation. Thus, parasites do not need genes that code for materials and functions provided by the host. Examples of such host-dependent, degenerate bacteria include the obligately intracellular parasites *Rickettsia* and *Chlamydia*. *Chlamydia* species produce special elementary reproductive bodies in cells that can be as small as 0.2 μm in diameter, somewhat smaller than cells of *Rickettsia* spp.

Another small parasitic bacterium is *Bdellovibrio*, which has a typical Gram-negative cell wall. This Proteobacterium is about 0.25 μm in diameter and about 0.5 μm in length. It is a parasite of other Gram-negative bacteria.

The mycoplasmas comprise yet another group of small parasitic bacteria. These organisms lack cell walls and may be as small as about 0.2 to 0.3 μm in diameter. It is noteworthy that the mycoplasmas are all host-dependent parasites and pathogens, so they would typically be found associated with larger host organisms. It is much more likely that the host would leave a fossil record than these cell wall-less bacteria.

Oligotrophy

Many natural aquatic and soil environments, such as the pelagic marine water column, have very low concentrations of nutrients. Living in these environments are oligotrophic bacteria that select for organisms with high surface area to volume ratios (SA/V) to enhance nutrient uptake. Because the environment is nutrient limited, oligotrophic bacteria do not need to grow rapidly and therefore do not need to produce large numbers of ribosomes and enzymes. Thus, small organisms that have a high SA/V and few ribosomes and enzymes have a selective advantage in such environments.

Production of Small Reproductive Cells and Spores

Most bacteria divide by binary transverse fission. In this process two cells of comparable size and mirror-image symmetry are produced. The daughter cells receive about half of the material and energy of the parent cell, and the cell diameter remains unchanged throughout the division cycle. One possible strategy for reproduction would be to produce a small reproductive cell that would have the minimal requirements for independent growth. The mother cell in this instance would not commit so much of its resources to reproduction as would be required if the daughter cell were the same size as the parent cell. Two examples of cell division processes, budding and baeocyte production, are known in bacteria that result in the production of cells that are smaller than the parent. In addition, some bacteria produce special hardy cells referred to as endospores, cysts, or exospores that may be smaller than the parent cell.

Buds and Baeocytes. Many bacteria produce buds. Examples of budding bacteria are reported in the phylogenetic groups Proteobacteria (e.g., *Hyphomicrobium*, *Prosthecomicrobium*, *Ancalomicrobium*, *Gemmiger*, etc.) and Planctomycetes (*Pirellula*, *Planctomyces*, *Gemmata*, and *Isosphaera*). However, in all groups reported above, the cell size of the mother cells is quite large, so although the bud diameters are smaller, they are still greater than 0.2 µm in diameter when they separate from their mother cells (*Bergey's Manual*, 1989).

Some Pleurocapsaen cyanobacteria undergo multiple fission to produce small cells referred to as baeocytes. However, those that have been reported are more than 1.0 µm in diameter (Waterbury and Stanier, 1978).

Endospores, Cysts, and Exospores. Endospores are special survival cells produced by some Gram-positive bacteria, particularly those that live in sediments, soil, and rock environments. The classical genera *Bacillus* and *Clostridium* are best known for endospore production, but others such as *Sporobacillus* also are known. The endospore contains DNA, ribosomes, and several layers of wall material referred to as a coat. The mature endospore is dehydrated and contains high concentrations of calcium and dipicolinic acid. Usually the endospore is somewhat smaller in diameter than its vegetative mother cell, but in some cases, such as *Clostridium tetani* (which causes tetanus), it is actually larger. However, none of the endospores reported is less than 0.25 µm in diameter (*Bergey's Manual*, 1986).

Cysts are produced as resting stages by some Gram-negative bacteria found in soils. *Azotobacter* species are one example. The myxobacteria also produce cysts termed microcysts or microspores. Cysts of both of these Proteobacterial groups are relatively large, ultimately larger than 0.25 µm, because they are formed from a vegetative cell by the addition of extra layers outside the cell wall.

Exospores or conidiospores are produced by many of high mol% G + C Gram-positive bacteria such as *Streptomyces* spp. These specialized cells are produced in the aerial mycelium as a resistant dispersal reproductive cell. They are about the same diameter as the filament diameter, greater than 0.5 µm (*Bergey's Manual*, 1989).

Other Small Free-living Organisms

A recently discovered small bacterium is a member of the division Verrucomicrobia, one of the major, more recently described phylogenetic groups of the Bacteria (Hedlund et al., 1996). This anaerobic free-living bacterium is about 0.35 μm in diameter and 0.5 μm in length (Janssen et al., 1997). *Thermoplasma* is an example of a small (0.2 μm diameter), free-living, cell-wall-less archaeon that is found in natural environments.

Many bacteria form very thin filaments. The spirochetes are one group that contains species whose cell diameters may be 0.1 μm. However, the cells are much longer, in excess of 5 μm (*Bergey's Manual*, 1984), so the minimum cell volume is comparable to that of cocci and rods.

Small Subcellular Structures

Small structures have the potential of producing small fossils, although this author is not aware that any of them have been reported as fossils. Candidate structures from contemporary bacteria include prosthecae and stalks that are extensions of the cell and that are smaller than the diameter of the cell. In addition, gas vesicles are very small proteinaceous structures formed by some Bacteria and Archaea. These structures are normally associated with the much larger organism that produces them. However, it is possible that, under some environmental conditions, they could be released from the parent cell and therefore become fossilized in its absence.

Prosthecae

Certain bacteria produce cellular appendages. Those of *Caulobacter* and *Asticcacaulis* may be quite narrow, approximately 0.1 μm in diameter. Furthermore, under some conditions, these structures can be separated from the cells giving rise to very thin membrane-bound structures that might be mistaken for cells. However, these structures would not be viable and would be expected to occur only rarely in natural environments. The prosthecae of *Hyphomicrobium*, *Pedomicrobium*, *Ancalomicrobium*, and *Rhodomicrobium* are about 0.2 μm in diameter and are less likely to become detached from the cell (Perry and Staley, 1997).

Stalks

Stalks are non-cellular appendages found on some bacteria such as *Gallionella* and *Planctomyces* spp. These structures may become encrusted with iron and manganese oxides. *Planctomyces* stalks are fibrillar consisting of several pilus-sized fibers several μm in length that are held together in a fascicle. They are often so fine, less than 0.1 μm in diameter, that they cannot be observed by light microscopy. However, *Gallionella* stalks may be much larger and because of encrustation may produce readily observable fossils in excess of 1.0 μm in diameter and up to several microns in length.

Gas Vesicles

Gas vesicles are proteinaceous membranes that are produced by many Bacteria and some Archaea. These structures are elongated cylinders with conical tips. They range in diameter from 45 to 200 nm and in length from 100 to more than 800 nm (Walsby, 1994). They are most abundantly produced by cyanobacteria during summer blooms in lakes, but are also produced by some heterotrophic bacteria and halophilic and methanogenic Archaea. Cyanobacterial cells may lyse at the end of a bloom releasing vesicles into the environment where they could become fossilized.

"Metallogenium"

One of the major findings in microbiology in the 20th century was the discovery of microbial fossils. The research of micropaleontologists, Barghoorn and Tyler (1965), revolutionized our thinking. The filamentous fossilized microstructures they found were so compellingly reminiscent of modern day cyanobacteria that their discovery convinced a whole generation of skeptical microbiologists about the existence of microbial fossils.

One of the major difficulties in studying ancient microbial fossils on Earth is that their predicted simple structure makes them difficult to identify. Therefore, we would predict that the first microorganisms would have been unicellular and may have lacked the typical cell wall structure of modern-day Bacteria and Archaea. Fossils of single unicellular bacteria might be very difficult to identify as biological structures. However, fossilized pairs (as cells formed during binary transverse fission) might be more readily recognized as being biological. In any event, fossil hunting in early sedimentary rocks on Earth poses special problems owing not only to the great age of the material, but also to the expected simplicity of the earliest organisms.

Most of the readily recognizable microbial fossils date from 1.0 to about 2.5 Ga bp. Convincing fossils of more ancient microorganisms are not so readily found. One of the more common precambrian fossils closely resembles modern microbial structures that have been named "Metallogenium" (Crerar et al., 1980). However, critical studies that have analyzed the modern-day counterpart that is found in the hypolimnion of lakes have cast doubt on its bacterial nature and/or viability (Klaveness, 1977; Gregory et al.,1980). Nonetheless, the possibility exists that the structure may be formed by microbial activities even though it is not a microorganism itself (Maki et al., 1987). This is an important point to verify in continuing research because, if this is true, its presence in fossilized material would be a signature of microbial life.

Acknowledgments

I appreciate the support of the National Science Foundation and the helpful comments of Brian Hedlund.

References

Barghoorn, E.S., and S.A. Tyler. 1965. Microorganisms from the gunflint chert. *Science* **147**:563-577.

Bergey's Manual of Systematic Bacteriology. 1984-1989. Vol. I, II, III, and IV (J.G. Holt, N.R. Krieg, J.T. Staley, and S. Williams, eds.). Baltimore, MD: Williams and Wilkins.

Crerar, D.A., A.G. Fischer, and C.L. Plaza. 1980. *Metallogenium* and biogenic deposition of manganese from Precambrian to recent time. Pp. 285-303 in *Geology and Geochemistry of Manganese* (I.M. Varentsov and G. Grasselly, eds.), Vol. III. Stuttgart: Schweizerbart'scheVerlag.

Gregory, E., R.S. Perry, and J.T. Staley. 1980. Characterization, distribution and signficance of *Metallogenium* in Lake Washington. *Microbiol. Ecol.* **6**:125-140.

Hedlund, B., J.J. Gosink, and J.T. Staley. 1996. Phylogeny of *Prosthecobacter*, the fusiform caulobacters: Members of a recently discovered division of the Bacteria. *Int. J. System. Bacteriol.* **46**:960-966.

Holm-Hansen, O., W.H. Sutcliffe, and J. Sharp. 1968. Measurement of deoxyribonucleic acid in the ocean and its ecological significance. *Limnol. Oceanogr.* **13**:507-514.

Janssen, P.H., A. Shuhmann, E. Mörschel, and F.A. Rainey. 1997. Novel anaerobic ultramicrobacteria belonging to the *Verrucomicrobiales* lineage of bacterial descent isolated by dilution culture from anoxic rice paddy soil. *Appl. Environ. Microbiol.* **63**:1382-1388.

Klaveness, D. 1977. Morphology, distribution and significance of the manganese-accumulating microorganism *Metallogenium* in lakes. *Hydrobiologia* **56**: 25-33.

Maki, J.S., B.M. Tebo, F.E. Palmer, K.H. Nealson, and J.T. Staley. 1987. The abundance and biological activity of manganese-oxidizing bacteria and *Metallogenium*-like morphotypes in Lake Washington, USA. *Microbiol. Ecol.* **45**:21-29.

Perry, J.J., and J.T. Staley. 1997. *Microbiology: Dynamics and Diversity.* Fort Worth, TX: Saunders College Publishing.

Walsby, A.E. 1994. Gas vesicles. *Microbiol. Rev.* **58**:94-144.

Waterbury, J.B., and R.Y. Stanier. 1978. Patterns of growth and development in Pleurocapsalean cyanobacteria. *Microbiol. Rev.* **42**:2-44.

SMALLEST CELL SIZES WITHIN HYPERTHERMOPHILIC ARCHAEA ("ARCHAEBACTERIA")

Karl O. Stetter
Lehrstuhl für Mikrobiologie, Universität Regensburg

Abstract

Hyperthermophilic archaea with optimal growth temperatures above 80° C represent the upper temperature border of life on Earth, occurring in volcanic and deep subterranean hot environments. Most of them are anaerobes able to use inorganic energy and carbon sources. Individual cells from pure cultures of members of the genera *Thermoproteus*, *Pyrobaculum*, *Thermofilum*, *Desulfurococcus*, *Staphylothermus*, *Thermodiscus*, *Pyrodictium*, *Thermococcus*, and *Pyrococcus* exhibit an exceptional variation in size. The volume of cells in the same culture may vary by more than four orders of magnitude. The smallest cell sizes observed in hyperthermophilic archaea are rods 0.17 µm in diameter in *Thermofilum*, spheres 0.3 µm in diameter protruding from rod-shaped cells of *Thermoproteus* and *Pyrobaculum*, and disks 0.2 to 0.3 µm in diameter and 0.08 to 0.1 µm wide in *Thermodiscus* and *Pyrodictium*. *Pyrodictium* forms web-like colonies in the centimeter range, in which the periplasmic space of the cells is connected to each other by a unique matrix of hollow tubules ("cannulae"). As a working hypothesis, the webs for the first time could enable an organism to use thermal gradients as an additional energy source. By their 16S rRNA-phylogeny, size-variable hyperthermophiles represent the shortest lineages closest to the root of the archaeal tree. Therefore, they may still be rather similar to their primitive ancestry at the early, much hotter Earth. The inability to keep their cell volumes constant may be seen as a primitive feature. However, by forming extremely small cells these organisms could be able to pass even pores of rocks in order to colonize deep subterranean environments.

Introduction

The first traces of life on Earth date back to the early Archaean age (Schopf, 1993; Mojzsis et al., 1996). Possibly, life had already existed about 3.9 billion years ago. At that time, there should have been an overall reducing atmosphere and a much stronger volcanism than today (Ernst, 1983). In addition, Earth's oceans were continuously heated by heavy impacts of meteorites. Therefore, within that scenario, early life had to be heat resistant to survive.

During the last decades, hyperthermophilic archaea had been isolated, which grow optimally (fastest) above 80° C, some even above 100° C (Stetter et al., 1981; Zillig et al., 1981; Stetter, 1982; Stetter and Zillig, 1985; Stetter, 1986; Stetter, 1996). Depending on the isolates, their minimum growth temperature is between 45 and 90° C, while their upper temperature border of growth is between 85 and 113° C (Table 1). Cultures of *Pyrolobus* and *Pyrodictium*, for the first time are even able to survive one hour autoclaving at 121° C, a kind of simulated "cosmic impact" scenario (Blöchl et al., 1997). Biotopes of hyperthermophiles are water-containing volcanic areas like terrestrial solfataric fields and hot springs, submarine hydrothermal systems, sea mounts, and abyssal hot vents ("Black Smokers"). The first evidence for the presence of communities of hyperthermophiles within geothermally heated subterranean rocks 3,500 meters below the surface of North Alaska was demonstrated recently (Stetter et al., 1993). Hyperthermophiles are well adapted to their biotopes, being able to grow at extremes of pH, redox potential, and salinity (see Table 1). Terrestrial hyperthermophiles usually require low salinity, while

Table 1 Growth Conditions of Some Hyperthermophilic Archaea

Species	Min. Temp. (°C)	Opt. Temp. (°C)	Max. Temp. (°C)	PH	Nutrition Autotrophic (a) Heterotrophic (h)	Biotope Submarine (s) Terrestrial (t)	Aerobic (ae) or Anaerobic (an)
Sulfolobus acidocaldarius	60	75	85	1-5	a/h	t	ae
Acidianus infernus	60	88	95	1.5-5	a/h	t/s	ae/an
Thermoproteus tenax	70	88	97	2.5-6	a/h	t	an
Pyrobaculum islandicum	74	100	103	5-7	a/h	t	an
Thermofilum pendens	70	88	95	4-6.5	h	t	an
Desulfurococcus mobilis	70	85	95	4.5-7	h	t	an
Staphylothermus marinus	65	92	98	4.5-8.5	h	s	an
Thermodiscus maritimus	75	88	98	5-7	h	s	an
Pyrodictium occultum	82	105	110	5-7	a/h	s	an
Pyrolobus fumarii	90	106	113	4.0-6.5	a	s	ae/an
Thermococcus celer	75	87	93	4-7	h	s	an
Pyrococcus furiosus	70	100	105	5-9	h	s	an
Archaeoglobus fulgidus	60	83	95	5.5-7.5	a/h	s	an
Ferroglobus placidus	65	85	95	6-8.5	a	s	an
Methanothermus sociabilis	65	88	97	5.5-7.5	a	t	an
Methanopyrus kandleri	84	98	110	5.5-7	a	s	an
Methanococcus igneus	45	88	91	5-7.5	a	s	an

those of marine biotopes are adapted to the high salinity of sea water. Most hyperthermophiles are strict anaerobes. A great many exhibit a chemolithoautotrophic mode of nutrition: inorganic redox reactions serve as energy sources, and CO_2 is the only carbon source required to build up organic cell material (Table 2). Depending on the organisms, hyperthermophiles are able to use H_2, ferrous iron, and reduced sulfur compounds as electron donors. On the other hand, oxidized sulfur compounds, nitrate, ferric iron, CO_2, and O_2 may serve as electron acceptors. Depending on the energy sources available, chemolithoautotrophic hyperthermophiles show great versatility: members of the same genera and even the same strains may be able to use different electron donors and acceptors (see Table 2). In addition, several hyperthermophilic archaea are facultative or obligate heterotrophs able to use organic compounds as

Table 2 Energy-yielding Reactions in Hyperthermophilic Archaea (Chemolithoautotrophes)

Energy-yielding Reaction	Genera
$FeS_2 + 7O_2 + 2H_2O \rightarrow 2FeSO_4 + 2H_2SO_4$	*Acidianus, Sulfolobus*
$2S^0 + 3O_2 + 2H_2O \rightarrow 2H_2SO_4$	*Sulfolobus, Acidianus*
$2H_2 + O_2 \rightarrow 2H_2O$	*Sulfolobus, Acidianus, Pyrolobus, Pyrobaculum*
$H_2S + HNO_3 \rightarrow HNO_2 + S^0 + H_2O$	*Ferroglobus*
$2FeCO_3 + HNO_3 + 5H_2O \rightarrow 2Fe(OH)_3 + HNO_2 + 2H_2CO_3$	*Ferroglobus*
$H_2 + HNO_3 \rightarrow HNO_2 + 2H_2O$	*Pyrolobus, Ferroglobus, Pyrobaculum*
$H_2 + S^0 \rightarrow H_2S$	*Acidianus, Pyrobaculum, Thermoproteus, Pyrodictium*
$4H_2 + H_2SO_4 \rightarrow H_2S + 4H_2O$	*Archaeoglobus*
$H_2 + 6Fe(OH)_3 \rightarrow 2Fe_3O_4 + 10H_2O$	*Pyrobaculum*
$4H_2 + CO_2 \rightarrow CH_4 + 2H_2O$	*Methanopyrus, Methanothermus, Methanococcus*

energy and carbon sources (see Table 1). Within the 16S rRNA-based phylogenetic tree, hyperthermophiles establish all the short and deep lineages (Figure 1; Woese et al., 1990). Short phylogenetic branches indicate a rather slow evolution. Therefore, by 16S rRNA phylogeny, hyperthermophiles represent the most primitive organisms known so far. The conclusion of thermophily as a primordial feature is in agreement with our picture of the early Earth. In this paper, I present results about variation and lowest limits of cell size within hyperthermophilic archaea

Morphology and Limits and Variation in Cell Size in Hyperthermophilic Archaea

In line with their great phylogenetic diversity, hyperthermophilic archaea display a variety of different cell morphologies (Table 3). Cells may be regular to irregular cocci, sometimes lobed or wedge-shaped, irregular disks with ultraflat areas, regular rods, or rods with spheres protruding at their ends ("golf clubs"). As usual for prokaryotes, cells in (pure) cultures of the euryarchaeotal *Methanothermus, Methanococcus,* and *Archaeoglobus* contain normal-sized rod-shaped or coccoid cells with not much variation in cell volume (see Table 3). The same is true for the coccoid-shaped *Sulfolobus* and *Acidianus* within the Crenarchaeota. A very special case is *Thermoplasma,* a thermoacidophilic heterotrophic cell-wall-less pleomorphic member of the Crenarchaeota. Cells are very flexible and propagate by budding. Cultures of *Thermoplasma* contain highly irregular cells with great variation in shape and size. The smallest cells observed are very tiny cocci, about 0.2 µm in diameter (see Table 3).

An unanticipated variation of cell sizes can be observed within pure cultures of members of the *Thermococcales, Desulfurococcales,* and *Thermoproteales,* which represent the deepest and shortest phylogenetic branches among the hyperthermophiles (Stetter, 1996): Cultures of *Thermococcus* and *Pyrococcus* usually show duplex-shaped irregular spheres, about 0.5 to 2 µm in diameter. However, during the early logarithmic growth phase, very tiny frog-egg-shaped cells about 0.2 µm in diameter arranged in clusters up to about 20 individuals may be observed. Sometimes, rather large cells show ribbon-like appendages that contain several very small cells in line (Stetter and Zillig, 1985). The function of these tiny cells is still unclear. However, after passing cultures of *Pyrococcus* through ultrafilters with 0.2 µm pore width, viable cultures could be obtained from the filtrates (Stetter, unpublished). Members of *Thermoproteus* and *Thermofilum* consist of stiff rectangular rods that show an extraordinary variation in length from about 1 to 100 µm. As a rule, during the exponential growth phase, tiny spheres about 0.3 to 0.5 µm in diameter are protruding at one end under an angle of 135°. Similar-sized spheres can be seen in cultures also in free state and may represent an unusual way of cell propagation. Alternatively, cells of *Thermoproteus* and *Thermofilum* are able to multiply by regular cell division. Strains of *Thermofilum* exhibit much thinner rod-shaped cells than *Thermoproteus*. Sometimes, cells of *Thermofilum* are only 0.15 to 0.17 µm in diameter, and therefore can hardly be recognized under the phase contrast light microscope, while cells of *Thermoproteus* consist of rather slim rods, about 0.4 µm in diameter (see Table 3).

Cultures of the heterotrophic *Staphylothermus* and *Desulfurococcus* reveal spherical cells with enormous variation in diameter between 0.5 and 15 µm. Therefore, their cell volume varies by more than four orders of magnitude. At low-nutrient concentrations morphology of cells of *Staphylothermus* is shifted mainly to giant cells, about 10 to 15 µm in diameter. The surface protein assembly of *Staphylothermus* (and possibly of the related *Desulfurococcus*) exhibits an unusual filamentous structure of extreme stability (Peters et al., 1995).

Cells of *Thermodiscus* consist of flat irregular disks, highly variable in diameter between 0.2 and 3 µm. They are about 0.1 to 0.2 µm wide. Sometimes pili-like structures of 0.01 µm in diameter and up to 15 µm in length connect the surfaces of two individuals. In the electron microscope, often extremely

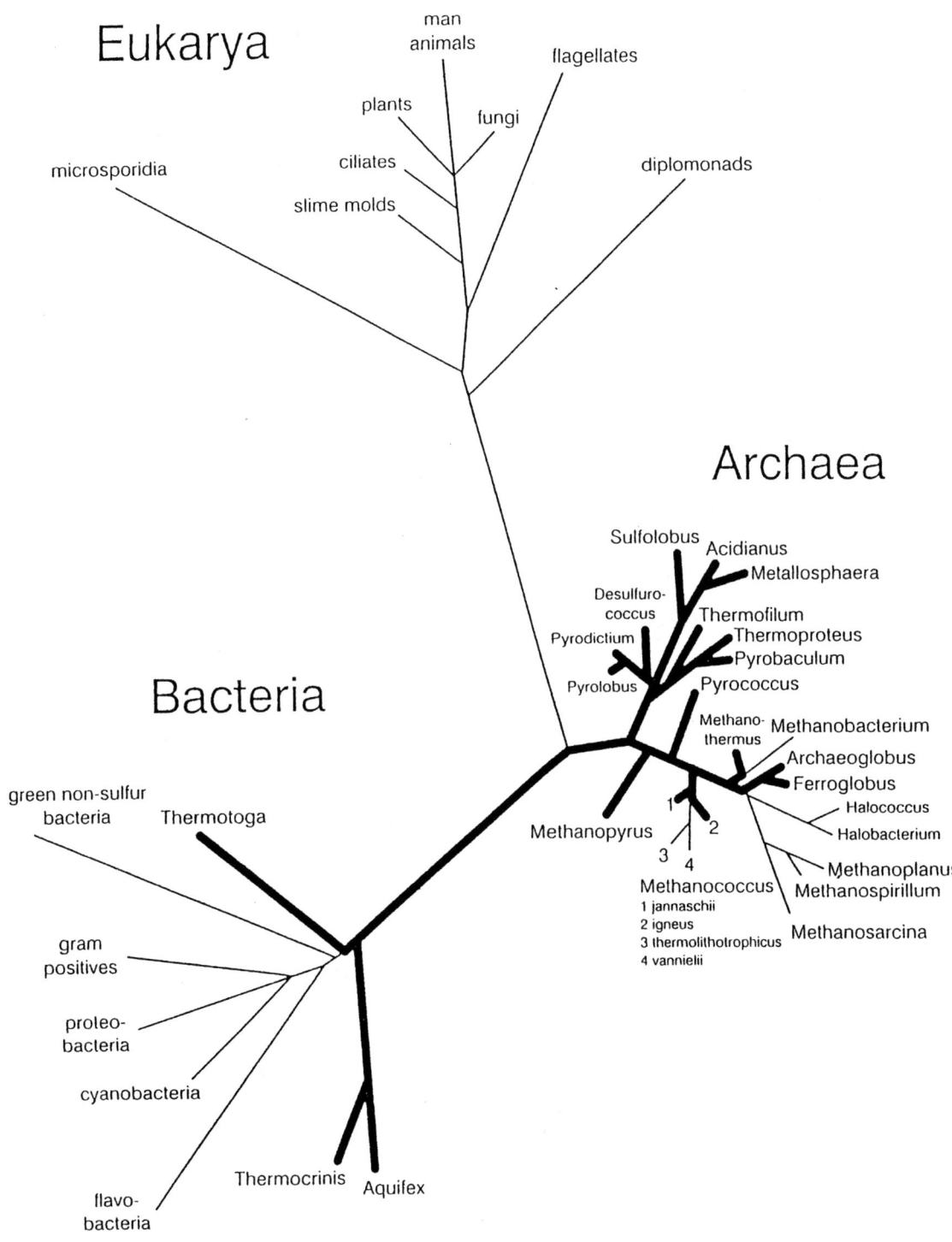

Figure 1. Hyperthermophiles (bulky lines) within the 16(18)S rRna-based phylogenetic tree.

Table 3 Morphology and Size of Hyperthermophilic Archaea (Examples)

Genus	Morphology	Size (μm)
EURYARCHAEOTA		
Archaeoglobus	Coccoid; wedge-shaped	0.4 - 1.0 dia.
Methanothermus	Rods	1 - 4 long; 0.3 - 0.4 dia.
Methanococcus	Spheres	1.3 - 2 dia.
*Thermoplasma**	Pleomorphic; buds	0.2 - 5 dia.
Pyrococcus	Spheres (+ "frog eggs")	0.5 - 2 dia.; (≈ 0.2 dia.)
CRENARCHAEOTA		
Sulfolobus	Lobed cocci	0.8 - 2 dia.
Acidianus	Lobed cocci	0.5 - 2 dia.
Thermoproteus	Branched rods + spheres ("golf clubs")	Rods: 0.4 dia.; 1 - 80 long; Spheres: 0.3 - 0.5 dia.
Thermofilum	Slender rods + spheres	Rods: 0.17 - 0.35 dia.; 1 - 100 long; Spheres: 0.3 - 0.5 dia.
Desulfurococcus	Spheres	0.5 - 15 dia.
Staphylothermus	Spheres in aggregates	0.5 - 15 dia.
Thermodiscus	Irregular disks	0.2 - 3 dia.; 0.1 - 0.2 wide
Pyrodictium	Irregular disks + matrix of cannulae	0.3 - 2.5 dia.; 0.08 - 0.3 wide; Cannulae: 0.026 dia.; up to 40 μm long

*Thermophilic, cell-wall-less archaeon.

small disks, less than 0.2 μm in diameter, are seen (Stetter and Zillig, 1985). This observation could explain that the titer as determined by serial dilution is always at least 10 times higher than that determined by direct counting in the light microscope.

Cells of *Pyrodictium* consist of flat, irregular disks. They are 0.3 to 2.5 μm in diameter and may be up to 0.3 μm wide. As a rule, cells of *Pyrodictium* exhibit large ultraflat areas, only about 0.08 μm in width (Rieger et al., 1997). Remarkably, *Pyrodictium* never grows in suspension but in mold-like flakes, several centimeters in diameter (Pley et al., 1991). The flakes are made up by a unique matrix of hollow tubules ("cannulae") in which the cells are integrated (Rieger et al., 1995). Single cannulae are up to 40 μm long and 0.026 μm in diameter and consist of glycoprotein subunits in helical array. The cannulae penetrate into the periplasmic space of the cells and connect those to each other, building up a huge network and greatly extending the range of a single cell. The flakes of *Pyrodictium* may be seen even as very primitve multicellular prokaryotic organisms. Because the cannulae represent a great deal of the biomass of *Pyrodictium* cultures, they should be of great importance. Five different structural cannulae genes identified so far do not show significant homology to any genes in other organisms including hyperthermophiles (Mai, 1998).

The advantage of the huge *Pyrodictium* web is still not evident and only a working hypothesis can be presented. In considering the physical uniqueness of the natural hot vent biotope and the great extension of cell range by the cannulae, for the first time in the living world *Pyrodictium* could be able to use thermal energy. *Pyrodictium* grows within the porous walls of deep sea "Black Smoker" vents, several millimeters to centimeters wide. From inside, the walls are strongly heated by the 300 to 400° C vent fluids, while they are cooled from outside by the surrounding 3° C deep sea water. Therefore, these chimneys harbour very steep temperature gradients, in which the *Pyrodictium* webs are situated, having a cold and a hot end. Interestingly, although this organism is growing only up to 110° C, its cannulae are stable up to 140° C. Similar to a thermocouple, electrons could be shifted in between the cold and hot end of the *Pyrodictium* web. This could cause changes in the membrane potential and finally ATP formation within the cells. At present, we are designing experiments to try this working hypothesis.

The enormous variation in cell size and volume appears to be a rather primitive feature and is in line with the 16S rRNA phylogeny of the corresponding hyperthermophiles. The smallest cell sizes observed are in the 200 to 300 nanometer range and the ability of hyperthermophilic archaea to form those may be of great advantage to pass narrow pores of soils and rocks in order to colonize hot subterranean environments.

References

Blöchl, E., Rachel, R., Burggraf, S., Hafenbradl, D., Jannasch, H.W., and Stetter, K.O. (1997). *Extremophiles* **1**, 14-21.

Ernst, W.G. (1983). The Early Earth and the Archaean rock record. Pp. 41-52 in *Earth's Earliest Biosphere, Its Origin and Evolution,* Schopf, J.W., ed. (Princeton University Press, Princeton, N.J.).

Mai, B. (1998). Thesis, University of Regensburg, Germany.

Mojzsis, S.J., Arrhenius, G., McKeegan, K.D., Harrison, T.M., Nutman, A.P., and Friends, C.R.L. (1996). *Nature* **384**, 55-59.

Peters, J., Nitsch, M., Kühlmorgen, B., Golbik, R., Lupas, A., Kellermann, J., Engelhard, H., Pfander, J.-P., Müller, S., Goldie, K., Engel, A., Stetter, K.O., and Baumeister, W. (1995). *J. Mol. Biol.* **245**, 385-401.

Pley, U., Schipka, J., Gambacorta, A., Jannasch, H.W., Fricke, H., Rachel, R., and Stetter, K.O. (1991). *System. Appl. Microbiol.* **14**, 245-253.

Rieger, G., Rachel, R., Hermann, R., and Stetter, K.O. (1995). *J. Struct. Biol.* **115**, 78-87.

Rieger, G., Müller, K., Hermann, R., Stetter, K.O., and Rachel, R. (1997). *Arch. Microbiol.* **168**, 373-379.

Schopf, J.W. (1993). *Science* **260**, 640-646.

Stetter, K.O. (1982). *Nature* **300**, 258-260.

Stetter, K.O. (1986). Diversity of extremely thermophilic archaebacteria. Pp. 39-74 in *Thermophiles: General, Molecular and Applied Microbiology*, Brock, T.D., ed. (John Wiley & Sons, Inc., New York).

Stetter, K.O. (1996). *FEMS Microbiol. Rev.* **18**, 149-158.

Stetter, K.O., and Zillig, W. (1985). Thermoplasma and the thermophilic sulfur-dependent archaebacteria. Pp. 85-170 in *The Bacteria,* Vol. III, Woese, C.R. and Wolfe, R.S., eds. (Academic Press Inc., Orlando).

Stetter, K.O., Huber, R., Blöchl, E., Kurr, M., Eden, R.D., Fielder, M., Cash, H., and Vance, I. (1993). *Nature* **365**, 743-745.

Stetter, K.O., Thomm, M., Winter, J., Wildgruber, G., Huber, H., Zillig, W., Janecovic, D., König, H., Palm, P., and Wunderl, S. (1981). *Zbl. Bakt. Hyg., I. Abt. Orig.* **C2**, 166-178.

Woese, C.R., Kandler, O., and Wheelis, M.L. (1990). *Proc. Natl. Acad. Sci. USA* **87**, 4576-4579.

Zillig, W., Stetter, K.O., Schäfer, W., Janekovic, D., Wunderl, S., Holz, I., and Palm, P. (1981). *Zbl. Bakt. Hyg., I. Abt. Orig.* **C2**, 205-227.

THE INFLUENCE OF ENVIRONMENT AND METABOLIC CAPACITY ON THE SIZE OF A MICROORGANISM

Michael W.W. Adams
Departments of Biochemistry and Molecular Biology
University of Georgia

Abstract

Simple calculations show that there are two critical factors in considering minimum cell size: the amount of DNA that is required to support cell growth and the volume of the cell devoted to accommodate that DNA. The amount of DNA a cell contains is related to how much that cell depends upon its environment to supply nutrients. At one extreme, the environment provides only gases and minerals, and the life-forms that occupy such an environment have a high biosynthetic capacity and synthesize all cellular carbon from CO_2. This requires at most 1,500 (an actual value) and perhaps as few as 750 genes. At the other extreme are nutrient-rich environments, such as those experienced by parasitic bacteria, and here life-forms have a minimum biosynthetic capacity requiring between 250 (a calculated value) and 500 genes (an actual value). For spherical cell with minimal biosynthetic capacity (250 genes), the minimum size is 172 nm diameter. This assumes that the cell consists (by volume) of 10% DNA, 10% ribosomes, 20% protein, and 50% water. Such a cell could contain 65 ribosomes and an average of 65 proteins per gene. On the other hand, a cell that synthesizes all of its cellular components from CO_2 must be at least 248 nm in diameter, assuming that its minimal DNA content (750 genes) is 10% of the cell volume. It is concluded that microorganisms cannot have diameters less than 172 nm if they have the same basic biochemical requirements for growth as all other extant life-forms. Even then, such a cell is biosynthetically challenged and would require a very specialized environment to supply it with a range of complex biological compounds. More likely, the absolute minimum size is closer to 250 nm where the cell has sufficient DNA to enable it to grow on simple compounds commonly found in various natural environments including, possibly, extraterrestrial ones.

Introduction

The question of minimum microbial size was recently brought to the fore by the report of McKay and coworkers (1) in which objects with upper dimensions of 20 by 100 nm were postulated to be of cellular origin. Subsequently, so-called ultramicrobacteria were isolated from marine environments that can pass through a 200 nm filter and have cell volumes of 0.03 to 0.08 μm^3 (2). In addition, entities known as "nanobacteria" that have been cultured from blood apparently have diameters as low as 80 nm (3). In light of these studies, it is important to estimate the theoretical limit for minimum cell size. Can cells with a diameter of less than, say, 50 nm contain sufficient biological material to remain free-living? This begs the question of what is meant by "sufficient biological material"? One measure is genome size or more specifically the number of different types of proteins (enzymes) that an organism has at its disposal to support growth. Before considering just how many genes this may be, we must also define what is meant by "free-living." How dependent is this minimally sized cell upon its environment? In the following it is assumed that such cells have the same basic biochemical requirements as any other life-form that we know of, and must satisfy them with the same enzymatic reactions.

The Influence of Environment on the Complexity of Life-forms

Obviously, no life-form survives in isolation from its surroundings, but organisms vary considerably in their dependence upon their environment. Thus, humans require at a minimum ten or so amino acids, various minerals, an array of biological cofactors (vitamins), and a continual supply of O_2 gas. Perhaps surprisingly, these same materials are also required by many microorganisms, although they typically differ from us in their ability to synthesize most, if not all, of the twenty amino acids as well as many, if not all, of what we term "vitamins." Like us, the vast majority of microorganisms require a fixed carbon source, which is usually a carbohydrate of some sort, although in some cases, lipids, nucleotides, or various simple organic compounds are utilized. In contrast, some microorganisms are intensely dependent upon their environments. For example, some microbial parasites do not synthesize any amino acid or lipid, and only a few enzyme cofactors and nucleotides; rather, they obtain all of these compounds from their host. Indeed, at one level, such a parasitic life-form is not too far removed from the simplest virus. This consists of a protein coat that surrounds a defined amount of nucleic acid (DNA or RNA). The latter encodes proteins that inside the host cell are synthesized and that direct viral replication. Hence a virus can be thought of as a life-form that has an extreme dependence upon its host. Not only does the host donate all of the necessary biological compounds, but it also provides transcriptional and translational machinery. What then is a plasmid? Can this be considered an extremely parasitic life-form? A plasmid obviously encodes the information to reproduce itself, i.e., to make a copy, but it is totally dependent upon the host to carry this out.

Hence, we can consider plasmids, viruses, and parasitic bacteria as life-forms that vary in their dependence upon their environments. So what forms of life are at the other extreme? What life-form requires the least from its environment? Obviously, these are organisms that require nothing more than the simplest of chemicals, such as CO_2, O_2, H_2, and NH_3. These so-called autotrophic organisms can synthesize all amino acids, cofactors, nucleotides, etc., with CO_2 as the sole carbon source, using the oxidation of H_2 as an energy source, and with ammonia (or even N_2 gas) as the nitrogen source. Interestingly, this definition also includes green plants—with the exception that they obtain energy from visible light rather than from H_2 oxidation. Of course, many autotrophic microorganisms gain energy from the oxidation with O_2 of simple substances other than H_2, such as CO, CH_4, NH_3, or H_2S. Similarly, anaerobic autotrophs growing on H_2 and CO_2 also conserve energy during the reduction of CO_2, either by the production of methane or acetate as accomplished by methanogens and acetogens, respectively.

Clearly then, the variety of known life-forms can be classified by the extent to which they depend upon their environment for growth. Simple gases and salts are sufficient for many types of microorganism, both under aerobic and anaerobic conditions, whereas other microbes are intensely dependent upon their environment for a range of complex biological molecules. So how do we define "free-living" life? In simple terms, life can be thought of as an entity that has the ability to undergo self-directed reproduction when supplied with the appropriate environment and the necessary free energy. The question is, can this environment be another life-form? If this is the case, then the argument becomes how small can a virus be, and a possible answer is a plasmid. However, an important difference between viruses (plasmids) and parasitic bacteria is that the former, but not the latter, replicate by the transfer of nucleic acid into their environment (host). With the bacteria, the host environment provides "only" an array of nutrients, and the bacterium's genetic material does not contact the host (the environment). Theoretically and often practically, the parasite can thrive if such nutrients are provided to it directly in a liquid medium. Hence, a major distinction can be made between the mechanisms by which parasitic life-forms and viruses interact with their "living" environments. Moreover, we can use this logic to define the environment that will support our smallest possible life-form. That is, we will assume that if its environment is another life-form, then nutrients (or non-life-forms) can replace that life-form. In other words, we will not consider viruses or analogous life-forms in trying to define minimum size.

Life-forms therefore occupy environments that fall between two extremes. One provides only gases and minerals, and the life-forms that occupy it must have a high biosynthetic capacity. At the other extreme are nutrient-rich environments, such as those experienced by parasitic bacteria, and here life-forms can have a minimum biosynthetic capacity. So, how many types of proteins (enzymes) are required to support cellular growth within these two types of environment? The recent availability of genome sequences for a variety of microorganisms (4) enables quantitative estimations to be made.

Organisms with Low Biosynthetic Capacity

Those organisms that are most dependent upon their environment are the parasitic bacteria, the prototypical example of which are the mycoplasma. Interestingly, the complete genome of one species, *Mycoplasma genitalium*, was one of the first genomes to be sequenced (5). At 0.58 Mb, this represents the smallest known genome of any free-living organism. The genome contains 470 predicted protein coding regions, and these include those required for DNA replication, transcription and translation, DNA repair, cellular transport, and energy metabolism. However, comparisons with the genome (1.83 Mb, encoding 1,703 putative proteins) of another parasite, *Haemophilus influenzae* (6), led to the conclusion that the "minimal gene set that is necessary and sufficient to sustain the existence of a modern-type cell" is (only) 256 genes, or about half of the genome of *M. genitalium* (7). It should be noted, however, that while both of these parasitic organisms grow in the absence of their hosts, to do so they require an extremely "rich medium" containing a range of nutrients. These organisms maintain a minimal biosynthetic capacity, a capacity that is apparently satisfied by approximately 250 different proteins.

The Most Slowly Evolving Microorganisms

In determining the "minimum" set of genes that a minimal-size microbe might contain, we must also consider what is meant by the term "modern-type cell" quoted above (7). Are present-day organisms highly sophisticated cells with a range of metabolic capabilities, only some of which are utilized and then under very specialized conditions? For example, *E. coli* could be regarded as highly evolved because it exhibits a range of metabolic modes, including growth under aerobic and anaerobic conditions, the utilization of a wide variety of different carbon sources, etc. Indeed, such a large metabolic capacity might be reflected in its genetic content of 4.64 Mb encoding 4,288 genes (8). Similarly, metabolically diverse species such as *Bacillus subtilis* and *Pseudomonas putida* have genomes of comparable size ($\geq$ 4 Mb). Indeed, a recent survey of gram-negative bacteria gave a mean genome size of 3.8 Mb (9). In other words, it is not unusual for microorganisms, or at least those that have been well characterized, to contain 4,000 or more genes. Hence, are there more-slowly-evolving organisms, and do they contain less genetic material and have fewer metabolic choices?

By phylogenetic analyses based on 16S rRNA sequence comparisons, the most-slowly-evolving microorganisms are the deepest branches, the first to have diverged within the two major lineages corresponding to the Bacteria and the Archaea (10). Remarkably, in both domains these are the hyperthermophiles, organisms that grow optimally at temperatures of 80°C and above. Within the bacteria domain this includes two genera, *Thermotoga* and *Aquifex,* while there are almost twenty genera of hyperthermophilic archaea (11). In fact, one of the two major branches within the archaeal domain consists almost entirely of hyperthermophiles, while in the other branch the hyperthermophiles are the most slowly evolving of the known genera. A great deal is known about the genome contents of these hyperthermophilic organisms as several have been or are being sequenced. These include the genomes of the archaea, *Methanococcus jannaschii, Pyrobaculum aerophilum, Pyrococcus horikoshii, P. furiosus, P. abyssi,* and *Archaeoglobus fulgidus,* and of the bacterium *Thermotoga maritima* (4). Interestingly, all of these organisms have genomes only about half the size of that found in *E. coli,* with those of *Archaeoglobus fulgidus* and *Aquifex aeolicus* being the largest (2.18 Mb) and smallest (1.55 Mb) of this group, respectively. Thus, the most slowly evolving organisms known (at least as determined by 16S rRNA analyses) do indeed have relatively small genomes, although they are still highly complex life-forms.

Organisms with High Biosynthetic Capacity

So, how many different proteins are required to support growth of organisms on nothing more than gases and a few minerals, and is there a hyperthermophilic example of such an organism? To date, the genomes of two hyperthermophilic autotrophs have been sequenced. One is the archaeon, *Methanococcus jannaschii* (12), which is a methanogen that grows up to 90°C using H_2 and CO_2 as energy and carbon sources and generates methane as an end product. The other is a bacterium, *Aquifex aeolicus* (13), which grows up to 95°C on H_2 and CO_2, but it is not an anaerobe like the methanogen, as it requires low concentrations of O_2. The genome sizes and number of proposed protein-encoding genes in these two organisms are 1.67 and 1.55 Mb, and 1,738 and 1,512, respectively. It should be noted that the pathway of CO_2 assimilation and the biochemistry of energy conservation in the methanogen are very different from those in *A. aeolicus,* yet approximately the same number of genes are required. On the other hand, these genomes are much larger than the genomes of the two parasites discussed above. Presumably, *A. aeolicus* and *M. jannaschii* require many more genes because they must synthesize all of their cellular components from CO_2. Hence they contain about three times the genetic information of *M. genitalium.* This seems appropriate considering that the latter organism is supplied with all of its amino acids, nucleotides, fatty acids, "vitamins," and with an energy source (glucose). From this direct comparison we might conclude that about two-thirds of the DNA in *A. aeolicus* and *M. jannaschii,* or approximately 1,000 genes, encodes proteins that function to carry out these biosynthetic tasks and produce all of these compounds from CO_2.

The Smallest Cell

From the above discussion it can be concluded that a cell with minimal biosynthetic capacity that is growing in a nutrient-rich medium requires between 250 (a calculated value) and 500 genes (the approximate number in *M. genitalium*) to grow. At the other extreme is a cell that synthesizes all of its cellular material from CO_2, and this requires at most 1,500 genes (the approximate number in *A. aeolicus*) and probably closer to 750 genes (half of the actual value). With these values in mind, let us consider how much biological material can be contained within a cell of, say, 50 nm diameter. For example, if one allows 5 nm in thickness for a lipid bilayer, a spherical cell of 50 nm diameter would have an internal volume of 33,500 nm^3. For comparison, an *E. coli* cell, with dimensions of about 1.3 by 4.0 μm, has an internal volume of about 5×10^9 nm^3, or almost 2 million times the volume of the 50 nm diameter cell. The question is, What quantities of the various biochemical structures found in a typical prokaryotic cell can be accommodated within a volume of 33,500 nm^3? A ribosome has a diameter of about 20 nm, and ribosomes are typically 25% of the mass (dry weight) of a bacterial cell (although this varies considerably depending on the growth rate). Assuming a similar percentage of the volume of a 50 nm diameter cell is devoted to them, such a cell could contain only two ribosomes (4,200 nm^3 each). Whether only two would limit cell growth to any extent is unknown; nevertheless, the cell is certainly large enough to contain ribosomes, albeit only two. On the other hand, proteins usually constitute about half of the dry weight of bacterial cells. Let us assume that they also occupy approximately 50% of the volume of the 50 nm diameter cell, and that, in general, proteins have an average molecular weight of 30 kDa, which corresponds to a diameter of about 4 nm per protein. If a cell of 50 nm diameter were 50% protein by volume, then this would correspond to about 520 such molecules (average 30 kDa) per cell.

Are two ribosomes and 520 "average-sized" protein molecules sufficient to support the growth of a cell? Note this would correspond to, on average, two copies of each protein for a cell with minimal biosynthetic capacity (calculated to contain 250 genes). But can we neglect DNA? As this is typically only about 3% of the total mass (dry weight) of a bacterial cell, at first glance it would seem unlikely that the volume of genetic material, especially in an organism with a minimum gene content, would affect cell size. For example, with a diameter of 2 nm and length of 0.34 nm/bp, the 4.64 Mb of *E. coli* has a volume of 4.9×10^6 nm^3, which is less than 1% of the cell volume. Surprisingly, however, simple calculations show that DNA is a determining factor in much smaller cells. Thus, the hypothetical 50 nm diameter cell, 75% of which (by volume) is occupied by proteins and ribosomes, could contain, even if the remaining 25% of the cell were devoted to DNA, only 8 genes (of 1000 bp each)!

DNA Content Determines Cell Size

If a 50 nm cell can only reasonably accommodate 8 genes, the question is, What is the minimum cell size that could reasonably accommodate 250 genes (or 250 kb of DNA)? Remarkably, even if the cell were 50% DNA, such a cell would have a diameter of at least 110 nm. Assuming that half of the remaining volume (25%) is protein and half of that (12.5%) is occupied by ribosomes, the 110 nm cell could contain up to 4,000 protein molecules (average 30 kDa) or an average of 8 proteins per gene, together with 15 ribosomes. Of course, such a cell would have minimal biosynthetic capacity. A cell growing on CO_2 as its carbon source would need at least 750 genes which, if they occupied 50% of the total volume, would require a cell of 156 nm in diameter. Such a cell could also contain 12,400 protein molecules (25% by volume, or 16 copies for each gene) and 48 ribosomes. Although such calculations still leave 12.5% of the cell volume for other cellular components, such as lipids, cofactors, metabolites, and inorganic compounds, the most abundant component of a typical cell, namely water, is not included. Water typically occupies about 70% of a microbial cell, so let us assume 50% for the hypothetical cell. The volume of a cell containing 250 genes then increases to 136 nm, while that with 750 genes is now 194 nm.

From these calculations it is obvious that DNA content is the main factor in determining cell size. A critical parameter is, therefore, the maximum amount of a cell that can be devoted to DNA. It seems extremely unlikely that DNA could represent 25% of the cellular volume (where water is 50%) if one considers just the volume occupied by the DNA molecule, with no allowance for neutralization of the negative charges, the bending of the DNA molecule, the unwinding of the double helix during replication and transcription, etc. It is hard to imagine that DNA could occupy more than 10% of the volume of a cell and still function. Thus, a cell that contains 250 genes that occupy 10% of its volume would be 172 nm in diameter, while one containing 750 genes occupying the same relative volume would be 248 nm in diameter. Assuming such cells contain 20% by volume protein and 10% by volume ribosomes, the 172 nm cell could accommodate 64 ribosomes and over 16,000 proteins, or 65 per gene, and the 248 nm cell could contain three times as much.

It may be concluded that the minimum theoretical size for a cell is 172 nm diameter. To grow, such a cell must be supplied with (and must assimilate) all amino acids, fatty acids, nucleotides, cofactors, etc., because it would contain the minimum number of genes (250) and have a minimal biosynthetic capacity. The cell would have a 5 nm membrane but no cell wall. It would consist, by volume, of 10% DNA, 10% ribosomes, 20% protein, and 50% water, and would contain approximately 65 proteins per gene as well as 65 ribosomes. In comparison, a cell with a much higher biosynthetic capacity, such that it could synthesize all cellular components from CO_2, would be 248 nm in diameter, assuming that its DNA is also 10% of the cell volume. Note that these calculations assume a theoretical minimum gene content, which is about half of that present in known life-forms. The amount of DNA in a known autotrophic organism (approximately 1,500 genes in *A. aeolicus*) would require a cell of at least 314 nm diameter, assuming that it occupied 10% of the cell by volume. Hence, depending on the biosynthetic capacity of a cell, and the extent to which the calculated minimum gene content (7) is realistic, its minimum diameter is between 172 and 314 nm. Overall, one can conclude that microorganisms cannot have diameters significantly less than 200 nm if they have the same basic biochemical requirements for growth as all other extant life-forms, but even then they would require very specialized environments. More likely, the absolute minimum size is closer to 250 nm where the cell is able to grow on simple compounds commonly found in various natural environments.

Acknowledgments

I thank Juergen Wiegel, Kesen Ma, and Jim Holden for helpful discussions.

References

1. McKay D.S., Gibson E.K., Thomas-Keprta K.L., Vali H., Romanek C.S., Clemett S.J., Chillier X.D.F., Maechling C.R., Zare R.N. (1996) Search for past life on Mars—possible relic biogenic activity in Martian meteorite ALH84001. *Science* **273**, 924-930.
2. Eguchi M., Nishikawa T., Macdonald K., Cavicchioli R., Gottscha J.C., Kjelleberg S. (1996) Responses to stress and nutrient availability by the marine ultramicrobacterium *Sphingomonas* sp. strain RB2256. *Appl. Environ. Microbiol.* **62**, 1287-1294.
3. Kajander E.O., Kuronen I., Åkerman K., Pelttaari A., and Çiftçioglu N. (1997) Nanobacteria from blood, the smallest culturable autonomously replicating agent on Earth. *SPIE* **3111**, 420-428.
4. Doolittle R.F. (1998) Microbial genomes opened up. *Nature* **392**, 339-342.
5. Fraser C.M., Gocayne J.D., White O., Adams M.D., Clayton R.A., Fleischmann R.D., Bult C.J., Kerlavage A.R., Sutton G., Kelley J.M., Fritchman J.L., Weiman J.F., Small K.V., Sandusky M., Fuhrmann J., Nguyen D., Utterback T.R., Saudek D.M., Phillips C.A., Merrick J.M., Tomb J.F., Dougherty B.A., Bott K.F., Hu P.C., Lucier T.S., Peterson S.N., Smith H.O., Hutchison C.A., Venter J.C. (1995) The minimal gene complement of *Mycoplasma genitalium*. *Science* **270**, 397-403.
6. Fleischmann R.D., Adams M.D., White O., Clayton R.A., Kirkness E.F., Kerlavage A.R., Bult C.J., Tomb J.F., Dougherty B.A., Merrick J.M., McKenney K., Sutton G., Fitzhugh W., Fields C., Gocayne J.D., Scott J., Shirley R., Liu L.I., Glodek A., Kelley J.M., Weidman J.F., Phillips C.A., Spriggs T., Hedblom E., Cotton M.D., Utterback T.R., Hanna M.C., Nguyen D.T., Saudek D.M., Brandon R.C., Fine L.D., Fritchman J.L., Fuhrmann J.L., Geoghagen N.S.M., Gnehm C.L., McDonald L.A., Small K.V., Fraser C.M., Smith H.O., Venter J.C. (1995) Whole genome random sequencing and assembly of *Haemophilus influenzae* RD. *Science* **269**, 496-512.
7. Mushegian A.R., Koonin E.V. (1996) A minimal gene set for cellular life derived by comparision of complete bacterial genomes. *Proc. Natl. Acad. Sci. USA* **93**, 10268-10273.

8. Blattner F.R., Plunkett G., Bloch C.A., Perna N.T., Burland V., Riley M., ColladoVides J., Glasner J.D., Rode C.K., Mayhew G.F., Gregor J., Davis N.W., Kirkpatrick H.A., Goeden M.A., Rose D.J., Mau B., Shao Y. (1997) The complete genome sequence of *Escherichia coli* K-12. *Science* **277**, 1453-1462.
9. Trevors J.T. (1996) Genome size in bacteria. *Antonie van Leeuwenhoek* **69**, 293-303.
10. Woese C.R., Kandler O., and Wheelis M.L. (1990) Towards a natural system of organisms: proposal for the domains of Archaea, Bacteria and Eucarya. *Proc. Natl. Acad. Sci. USA* **87**, 4576-4579.
11. Stetter, K.O. (1996) Hyperthermophilic prokaryotes. *FEMS Microbiol. Rev.* **18**, 149-158.
12. Bult C.J., White O., Olsen G.J., Zhou L.X., Fleischmann R.D., Sutton G.G., Blake J.A., Fitzgerald L.M., Clayton R.A., Gocayne J.D., Kerlavage A.R., Dougherty B.A., Tomb J.F., Adams M.D., Reich C.I., Overbeek R., Kirkness E.F., Weinstock K.G., Merrick J.M., Glodek A., Scott J.L., Geoghagen N.S.M., Weidman J.F., Fuhrmann J.L., Nguyen D., Utterback T.R., Kelley J.M., Peterson J.D., Sadow P.W., Hanna M.C., Cotton M.D., Roberts K.M., Hurst M.A., Kaine B.P., Borodovsky M., Klenk H.P., Fraser C.M., Smith H.O., Woese C.R., Venter J.C. (1996) Complete genome sequence of the methanogenic archaeon *Mechanococcus jannaschii*. *Science* **273**, 1058-1073.
13. Deckert G., Warren P.V., Gaasterland T., Young W.G., Lenox A.L., Graham D.E., Overbeek R., Snead M.A., Keller M., Aujay M., Huber R., Feldma R.A., Short J.M., Olsen G.J., Swanson R.V. (1998) The complete genome of the hyperthermophilic bacterium *Aquifex aeolicus*. *Nature* **392**, 353-358.

DIMINUTIVE CELLS IN THE OCEANS—UNANSWERED QUESTIONS

Edward F. DeLong
Monterey Bay Aquarium Research Institute

Abstract

The marine environment harbors enormous numbers of viruses and prokaryotes, existing in complex communities that span a wide spectrum of biotopes, lifestyles, and size ranges. Many naturally occurring marine bacterioplankton are extremely small, some measuring < 0.3 µm in their largest dimension, having estimated biovolumes as low as 0.027 µm^3. Available data suggest that the majority of naturally occurring bacterioplankton resist cultivation, and have not been phylogenetically identified at the single cell level. Phylogenetic evidence for the evolution of major lineages that are characteristically small have not been reported (but they may exist). Because a large fraction of naturally occurring microorganisms have not been cultivated, their specific physiological traits are largely unknown. Consequently, the fraction of very small marine microbes that transiently and reversibly exist as "dwarf cells" is also unknown. Finally, although extremely small (< 0.1 µm) DNA-containing particles are very abundant in seawater and are thought to be viruses, the fraction of these particles that may actually represent cellular organisms is uncertain.

Introduction

Small microorganisms are ubiquitous in ocean waters, averaging about 5×10^5 cells/ml in the upper 200 m, and 5×10^4 cells/ml below 200 m depth. The total number of prokaryotic cells in ocean waters is about 1×10^{29}(1). Assuming a biomass of approximately 20 fg carbon per cell, this represents 2.2×10^{15} g of prokaryotic carbon in the world's oceans. This biomass represents an enormous pool of genetic variability, a large fraction of which is represented by very small cells (2,3). Extremely small cells (< 0.5 µm) may result from a genetically fixed phenotype maintained throughout the cell cycle. Alternatively, very small cells may reflect physiological changes associated with starvation, or other aspects of the cell cycle. Both explanations likely hold for different members of complex mixed populations of small cells found in the ocean. Extremely small (<0.1 µm) DNA-containing particles are also very abundant in seawater, reaching concentrations of about 1×10^7 particles/ml in surface waters (4-6). These small particles are thought to consist largely, although not necessarily entirely, of viruses.

Cell dimensions of cultured or naturally occurring bacteria can be derived from several sorts of data, each with inherent limitations. A number of uncertainties can be associated with cell size estimates. Historically, the existence of very small bacteria and viruses was first documented by observations of infectious filterable agents. Indirect cell size estimates have more recently been derived from filter fractionation experiments using membrane filters of uniform pore size. These sorts of size estimates can be compromised by filter trapping effects, as well as differential retention of cells with varying shapes or cell wall compositions. Cell dimensions and biovolumes are now more frequently estimated via fluorescent nucleic acid staining and epifluorescence microscopy, or flow cytometry. Fluorescent DNA stains can also sometimes be misleading, because the visualized nuclear material may not accurately reflect the actual cytoplasmic volume (7). Most estimates by light microscopy, electron microscopy, and flow cytometry also involve the use of fixatives that may cause cell shrinkage or other artifacts (3). Nevertheless, it is apparent that the majority of naturally occurring prokaryotes in marine plankton are about 1.0 μm or less in their largest dimension, and a good number of these are 0.5 mm or less in diameter (2,3).

Critical Assessment of the Issue

1. *What is the phylogenetic distribution of small bacteria?*

This question can be broken down into several components:

A. *What is the phylogenetic distribution of cultivated prokaryotes with a stable, very small cell size?*

The ongoing efforts of microbiologists to cultivate new microbial groups are currently providing new perspectives and answers to this question. It is still an open-ended question, because new microbial groups continue to yield to cultivation efforts. Recently isolated bacteria having stable, maximal dimensions of around 0.5 μm, fall into the alpha Proteobacterial lineage, as well as the Bacterial order Verrucomicrobiales.

Very small bacteria in the order Verrucomicrobiales have been recently isolated. New strains isolated from an anoxic rice paddy displayed a stable cell size of about 0.5 μm in length and 0.35 μm in diameter yielding a cell volume of about 0.03 μm^3 (8). These bacteria were oxygen-tolerant heterotrophs, exhibiting strictly fermentative growth on sugars. Other cultivated relatives, including *Verrucomicrobium spinosum*, are generally larger than 1 μm and possess prosthecae (9). Small cell size is therefore not an inherent property of members of this order.

A very small marine isolate with cell volume ranging from 0.03 to 0.07 μm^3 was isolated using the dilution culture technique of Button and Schut (10). This isolate was found to be associated with the alpha Proteobacterial genus *Sphingomonas* (11). *Sphingomonas* sp. strain RB2256 is heterotrophic, contains about 1.5 fg DNA/cell, and grows at a maximal rate of about 0.16 hr^{-1}. This marine *Sphingomonas* isolate showed very little variation in growth rate or cell size in response to 1,000-fold variation in nutrient supply, indicating the stability of the small cell phenotype (12). Other *Sphingomonas* species have larger, more typical cell sizes, so diminutive size is not a specific characteristic of the genus.

Nanobacteria species have been reportedly found in association with human and cow blood (13). They have been cultured in serum-free media, and have cell diameters, estimated from electron microscopy, of 0.2 to 0.5 μm (13). They have been reported to pass through 0.1 mm filters, apparently due to pleomorphic forms even smaller, about 0.05-0.2 μm (13). Ribosomal RNA sequences originating from these microorganisms are associated with the alpha subdivision of the Proteobacteria, and are most closely related to *Phyllobacterium rubiacearum*.

B. *What is the phylogenetic distribution of cultured prokaryotes that undergo an induced cell cycle transition from a "typical" to very small cell size?*

A significant number of bacteria have been observed to undergo a transition from a large, actively growing state, to a dormant state of much smaller cell size (14-16). Some of these physiologically induced small cells reduce to cell volumes as low as 0.03 μm^3. Different bacterial genera have been reported to undergo a starvation-induced response resulting in cell miniaturization, including the gamma Proteobacteria genera *Vibrio*, *Pseudomonas*, *Alcaligenes*, *Aeromonas*, and *Listonella* (14). This reduction in cell size may be a common phenomenon for heterotrophic microorganisms adapted for growth at relatively high nutrient concentrations. In many of these microorganisms, the transition from large to dwarf cells is fully reversible upon nutrient upshift. This physiological strategy appears to be common, but its actual distribution among diverse bacterial phyla is poorly characterized. It is unknown what fraction of naturally occurring "ultramicrobacteria" represent typically larger cells that have experienced nutrient downshift and undergone cellular miniaturization. It is also not clear what fraction of these readily reverse to a large actively growing state (15), or alternatively have entered a hypothetical "viable but nonculturable" state (16).

C. *What are the phylogenetic identities of (uncultivated) very small cells frequently observed in natural environmental samples?*

This remains an open question. It has been estimated that only about 0.1-1% of naturally occurring prokaryotes have been cultivated from many specific habitats (17,18). Culture independent surveys have indicated the presence of many new, yet uncultivated, and previously unrecognized prokaryotic groups (19). Most of these have not yet been specifically identified at the single cell level. It will be interesting to determine whether a significant fraction of recently discovered, uncultivated prokaryotic groups represent some of the more diminutive cell forms. Are there inherent properties of very small cell lineage that render them recalcitrant to cultivation?

2. *Is there a relationship between minimum size and environment?*

In low-nutrient habitats in marine plankton, cells typically appear smaller in size than those of comparable higher nutrient habitats. To the extent that some cells undergo a starvation response that involves reduction in cell size, there may be a loose relationship between cell size and ambient nutrient concentration. However, it is still unknown what fraction of naturally occurring small cells represent physiologically induced forms, versus stable, diminutive phenotypes. Smaller cells have a greater surface area to volume ratio, postulated to be adaptive for low-nutrient environments (11). However, small cell size does not necessarily imply adaptation to an oligotrophic (low-nutrient) lifestyle. For instance, new Verrucomicorbiales isolates (8) grow well and maintain small cell size under relatively high nutrient growth conditions (e.g., 4 mM glucose, or 0.1% starch). Nanobacteria dwell (and are cultivated) in a relatively nutrient-rich environment, yet maintain their small cell dimensions (13). Symbiotic and parasitic bacteria are known that have reduced physiological capacities and genome sizes (20). It is possible that symbionts in environments rich with host-supplied growth factors may actually have reduced genetic and physiological demands, thereby facilitating cell size reduction. It is possible that small cell size is adaptive for free-living cells in low nutrient environments, but symbiotic species may tend toward small cell size in a nutrient-replete environment provided by the host.

3. *Is there a continuum (or quanta?) of size and complexity that links conventional bacteria and viruses?*

Direct examination of concentrated seawater samples by electron microscopy have revealed the presence of large numbers of viral-like particles (VLPs) in the world's oceans (4,5). Ranging in numbers from about 2×10^5 to 5×10^8 particles/ml, VLP numbers typically exceed bacterial cell numbers in aquatic samples by 10-fold. Most quantitative studies to date have employed ultracentrifugation or ultrafiltration coupled with electron microscopy, or filtration, fluorescent DNA staining, and epifluorescence microscopy. A few studies have succeeded in enumerating naturally occurring viable infectious particles (especially in marine *Synechococcus* sp.) to determine the host range, in situ titers, and ecological variability of naturally occurring cyanophages (21).

In the marine environment there is certainly a continuum of size in both bacterioplankton and virioplankton. Bacterioplankton can range from large filaments > 10 µm, to small coccoid cells with diameters approaching 0.3 µm (2). Marine virus isolates range in length from about 40 nm to as large as 120 nm (5). Electron micrographs of naturally occurring infected cells suggest that some bacterial hosts are considerably less than 10-fold larger than their viral parasites, having a burst size of about 7 (6)! The very smallest bacterial cells and the very largest viral particles fall into about the same size category, raising some questions about the accuracy of currently used methods for quantifying naturally occurring virus and prokaryotes. Commonly used epifluorescence techniques are convenient and reproducible, but the identity of the fluorescently stained particles is certainly subject to some uncertainty. What fraction of VLPs are actually viruses? What fraction of VLPs are viable viruses? What fraction of DNA-containing particles < 0.1 µm are actually cells, and not viruses? If some of the < 0.1 µm DNA-containing particles are cells, are they viable? These remain open-ended questions.

With regard to the complexity of these populations, the issue of cultivability is a serious one. It still appears from available data that a large fraction of naturally occurring microbes have resisted cultivation attempts. The specific physiological traits and life histories of these microorganisms remain unknown, as does that of their viral parasites. A major challenge to contemporary microbiology is to devise and implement approaches to better characterize this large and uncharacterized biota.

References

1. Whitman, W.B., Coleman, D.C., Wiebe, W.J. (1998), *Proc. Natl. Acad. Sci. USA* **95**:6578-6583.
2. Watson, S.W., Novitsky, T.J., Quinby, H.L., Valois, F.W. (1977), *Appl. Environ. Microbiol.* **33**:940-946.
3. Fuhrman, J.A. (1981), *Mar. Ecol. Prog. Ser.* **5**:103-106.
4. Bergh, O., Borsheim, K.Y., Bratbak, G., Heidal, M. (1989), *Nature (London)* **340**:467-468.
5. Borsheim, K.Y. (1993), *FEMS Microb. Ecol.* **102**:141-159.
6. Steward, G.F., Smith, D.C., Azam, F. (1996), *Mar. Ecol. Prog. Ser.* **131**:287-300.
7. Suzuki, M.T., Sherr, E.B., Sherr, B.F. (1993), *Limnol. Oceanogr.* **38**:1566-1570.
8. Janssen, P.H., Schuhmann, A., Morschel, E., Rainey, F.A. (1997), *Appl. Environ. Microbiol.* **63**:1382-1388.
9. Hedlund, B.P., Gosnik, J.J., Staley, J.T. (1996), *Appl. Environ. Microbiol.* **46**:960-966.
10. Schut, F., DeVries, E.J., Gottschal, J.C., Robertson, B.R., Harder, W., Prins, R.A., Button, D.K. (1993), *Appl. Environ. Microbiol.* **59**:2150-2160.
11. Schut, F., Prins, R.A., Gottschal, J.C. (1997), *Aquat. Microb. Ecol.* **12**:177-202.
12. Eguchi, M., Nishikawa, T., MacDonald, K., Cavicchioli, R., Gottschal, J., Kjelleberg, S. (1996), *Appl. Environ. Microbiol.* **62**:1287-1294.
13. Kajander, E.O., Çiftçioglu, N. (1998), *Proc. Natl. Acad. Sci. USA* **95**:8274-8279.
14. MacDonnell, M.T., Hood, M.A. (1982), *Appl. Environ. Microbiol.* **43**:566-571.
15. Kjelleberg, S., Hermansson, M., Marden, P. (1987), *Ann. Rev. Microbiol.* **41**:25-49.
16. Rozak, D.B., Colwell, R.R. (1987), *Microbiol. Rev.* **51**:365-379.
17. Staley, J.T., Konopka, A. (1985), *Ann. Rev. Microbiol.* **39**:321-346.
18. Amann, R.I., Ludwig, W., Schleifer, K.H. (1995), *Microbiol. Rev.* **59**:143-169.
19. Pace, N.R. (1997), *Science* **276**:734-740.
20. Fraser et al. (1995), *Science* **270**:397-403.
21. Waterbury, J.B., Valois, F.W. (1993), *Appl. Environ. Microbiol.* **59**:3393-3399.

Panel 3

Can we understand the processes of fossilization and non-biological processes sufficiently well to differentiate fossils from artifacts in an extraterrestrial rock sample?

DISCUSSION

Summarized by Andrew Knoll, Panel Moderator

Recognition of a Biological Signature in Rocks

Dr. Knoll opened the session by summarizing the challenges of recognizing biological signatures in extremely old rock samples from Earth or Mars. Everyday experience suggests that the gulf between biology and the physical world is conspicuous. This impression arises, however, because the biology most familiar to us is principally that of organisms found on distal branches of the tree of life. The difficulty in distinguishing between biogenic and abiogenic features lies at the other end of the tree. Life arose as a self-perpetuating product of physical processes, and it is likely that the characteristics of Earth's earliest organisms—their size, shape, molecular composition, and catalytic properties—bore a close resemblance to the products of physical processes that gave rise to biology. For this reason, detecting the remnants of early life in terrestrial rocks is difficult. In martian or other extraterrestrial samples, it is doubly challenging. Given the evidence in hand from ALH84001 and the prospect of analyzing intelligently chosen samples from Mars within a decade, how do we fashion ground rules for recognizing the unambiguous signal of past (or present) biology?

Organisms have structure, they have a chemical composition, and they affect their environment; thus, paleontological evidence of ancient life can be morphological, geochemical, or sedimentological. Experience with terrestrial rocks makes it clear that features found in ancient samples can be accepted as biological only if they satisfy two criteria. They must be compatible with pattern generation by known biological processes. And they must be incompatible with formation by physical processes. This is straightforward in principle, but it requires that we understand the limits of pattern formation by biological and physical processes.

Dr. Knoll illustrated the challenge for this panel by drawing a Venn diagram in which biological and abiological patterns were depicted as distinct but overlapping fields. As yet, our understanding of the limits on both biological and abiological pattern generation is incomplete, making it difficult to understand the dimensions of the "gray zone" of overlap. Recognition of biological pattern in extraterrestrial samples will require the identification of structures or molecules that reside in the biological field, but not in the zone of overlap. On the other hand, there is no assurance that terrestrial life exhausts the possibilities of biological pattern generation; therefore, knowing the limits of pattern formation by physical processes may provide the best yardstick for evaluating martian or other extraterrestrial samples. Dr. Benner suggested an alternative depiction of the Venn diagram in which the biological field is completely encompassed within the physical field—the point being that biological processes are a subset of a larger and more inclusive set of physical processes. As a depiction of process, this view is unimpeachable; nevertheless, the patterns generated by biological processes include structures and molecules not known to form under strictly physical conditions. Bones, radiolaria, and red algal thalli are examples of biologically diagnostic morphologies; cholesterol is a biologically diagnostic molecule.

Lessons from Earth

Panel 3 members agreed that our collective experience with Earth's geological record provides an important guide to fossil recognition and interpretation on Mars. Dr. Farmer demonstrated that processes of mineral precipitation can preserve biologically interpretable microfossils and sedimentary fabrics. This increasing knowledge of fossilization processes not only sheds light on the postmortem information loss that attends fossilization, but also focuses attention on martian environments most likely to preserve a biological record. Preservation of terrestrial remains is selective, with some organisms—and some parts of organisms—more likely to escape decay than others. During fossilization, cells can also shrivel or collapse, resulting in fossils that are much smaller than the organisms from which they are derived.

Dr. Schopf summarized experience in interpreting Earth's early fossil record, stressing the early phase of discovery, when reports of objects that proved to be abiological outnumbered those of genuine fossils. Not everything that is small and round is biological, and the rigorous criteria for biogenicity developed over the past three decades by paleontologists can be useful in the evaluation of extraterrestrial microstructures. Dr. Schopf emphasized the need to conduct interdisciplinary studies of petrology, micropaleontology, isotopic geochemistry, and molecular organic geochemistry, developing multiple lines of evidence for interpreting potentially biological patterns.

Although most of Panel 3's discussion focused on micron-scale structures, the distinctive macroscopic structures known as stromatolites were also considered. Stromatolites are laminated structures found in chemical sedimentary rocks, especially but not exclusively carbonate rocks. These structures, which can be flat-laminated, domal, columnar, or conical, are commonly interpreted as the products of sediment trapping, binding, and/or precipitation by microbial communities; however, it is apparent that comparable structures can be generated without the need for microbial templates. Indeed, such structures actually occur in the early geological record. This being the case, images of laminated precipitates that may be transmitted by a Mars rover cannot be construed as unambiguous evidence for biological activity (Farmer, Knoll). Micro- and mesoscale fabric studies on returned samples will be necessary to confirm or reject hypotheses of biological origin. Dr. Farmer suggested that specific microscopic textures may provide the biological fingerprint needed to be confident of stromatolite biogenicity in the terrestrial or martian rock record.

Dr. Bradley focused attention on abiologial pattern formation, suggesting that non-biological processes may be sufficient to explain a number of micro- and nanoscale features sometimes interpreted as biological, including those reported from Mars meteorite ALH84001. Dr. McKay vigorously disputed some of these conclusions but agreed that much clearer criteria for biological pattern formation in extraterrestrial samples are needed.

Summary and Consensus

General consensus was reached on the following points:

• Terrestrial rocks contain an observable and interpretable record of biological evolution, but as we recede further back into time, that record becomes attenuated and difficult to interpret in detail. Martian samples may actually be better preserved than terrestrial sediments of comparable age, but lack both modern martian organisms for comparison and a more or less continuous fossil record that connects the present with early planetary history.

• A better understanding of biological signatures in sedimentary rocks is needed, and it is needed before intelligently collected martian samples are returned to Earth. These signatures certainly include fossil morphologies, but they must also include biomarker molecules, isotopic fractionation, and biological mineralization and trace element concentrations. In all cases, improved understanding of biological pattern formation must proceed in tandem with better knowledge of the generative capacity of physical processes.

• There is both a need and an opportunity to more effectively integrate laboratory and field observations of fossilization processes with investigations of Earth's early sedimentary record. Multi-disciplinary investigations are required in exopaleontological research, and there is a need for new technologies that will enhance our ability to obtain chemical information from individual microstructures.

FOSSILS AND PSEUDOFOSSILS:
LESSONS FROM THE HUNT FOR EARLY LIFE ON EARTH

J. William Schopf
Department of Earth and Space Sciences, Molecular Biology Institute,
IGPP Center for the Study of Evolution and the Origin of Life
University of California at Los Angeles

Abstract

As shown by other papers in this workshop, the actual and theoretical size limits of minute living microorganisms are incompletely defined and, so, the biogenicity of exceedingly small "biologic-like" objects is subject to debate. But the problem of distinguishing between minute biologic and nonbiologic objects is more vexing if they are present in rocks and therefore open to interpretation either as microbial fossils or nonbiologic pseudofossils, and it is a problem confronted even when interpreting objects with the dimensions of "normal-sized" microorganisms. The problem becomes more complicated still as the size and complexity of the objects considered decrease, or as the provenance of such objects becomes increasing distant from the present either in time (as for Archean in contrast with Proterozoic fossil-like objects) or space (as for structures in exterrestrial rather than terrestrial materials). Lessons learned from the past three decades of search for ancient records of life on Earth point a way toward solution of this problem: (1) The search for evidence of early life should be multidisciplinary (or, better, interdisciplinary); (2) the evidence should be positive, affirming a biologic orgin, rather than neutral (consistent either with biology or nonbiology) or negative (evidence inferred by default to be biogenic); and (3) the provenance, age, indigenousness, syngenicity, and biogenicity of such evidence must be established beyond question.

You must not fool yourself—and you are the easiest person [for you] *to fool.*

—Richard Feynman, CalTech
Commencement Address, 1974

Rules for the Hunt

There are fine lines between what is known, guessed, and hoped for, and because science is done by real people these lines are sometimes crossed. But science is not a guessing game. The goal is to know. "Possibly . . . perhaps . . . maybe" are not firm answers and feel-good solutions do not count. With regard to the famed Mars meteorite (McKay et al., 1996), for example, life either once existed on Mars or it didn't. Meteorite ALH84001 either holds telling evidence or it doesn't. Eventually, hard facts will sort it out. But the still-simmering controversy might have been avoided had three lessons now learned from the episode been heeded from the outset:

First, the search for evidence of past life, whether in Earth rocks or extraterrestrial samples, must be multidisciplinary—at a minimum, based on the techniques and findings of biologists, paleobiologists, geologists, and geochemists—or, better still, *inter*disciplinary, carried out by researchers schooled in both the life *and* physical sciences.

Second, the evidence sought should be *positive*—evidence that affirms the biologic origin of the features detected. Evidence that is *neutral* (consistent either with biology or nonbiology) is by its nature inadequate to establish the existence of past life, and interpretations based on *negative* reasoning—inference by default, such as a claim that because a feature is not obviously mineralic it "must" be biogenic—are likely to prove erroneous.

Third, to be acceptable, evidence of past life must meet five specific tests (Schopf and Walter, 1983; Schopf, 1993):

1. Provenance—Is the source firmly established of the rock sample that contains the putative biologic features? (For a terrestrial sample, stratigraphic and geographic provenance should be known precisely, as demonstrated, for example, by replicate sampling by different workers.)

2. Age—Is the age of the rock sample known with appropriate precision? (For example, is the age tightly constrained by multiple measurements and lines of evidence?)

3. Indigenousness—Are the putative biologic features indigenous to the rock? (For example, are the features embedded in the rock matrix rather than being surficial contaminants, a test that can be met for minerals and fossils by studies of petrographic thin sections and for biologic isotopic signatures by ion microprobe analyses in situ.)

4. Syngenicity—Are the features syngenetic with a primary mineral phase of the rock? (Or are they of later origin, for example introduced into pores or fractures and lithified by secondary or later-generation minerals—a test that for minerals, fossils, and isotopic evidence can usually be met by studies of petrographic thin sections.)

5. Biogenicity—Are the features assuredly biological? (Of the five tests, establishment of biogenicity has often proven to be the most difficult, as discussed below.)

The Search for Ancient Fossils

Four independent but potentially mutually reinforcing lines of evidence have been used to trace the earliest (microbial) records of life on Earth: minerals, organic compounds, isotopic signatures, and fossils. Of these, fossils are potentially the least ambiguous—independent of the other lines of evidence, the presence of unquestionable fossils indicates the existence of past life. On Earth, two kinds of fossils are present in the early rock record: stromatolites (a type of "trace fossil") and cellularly preserved microbes ("body fossils").

Stromatolites

Formally defined, a stromatolite is an accretionary organosedimentary structure, commonly thinly layered, megascopic, and calcareous, produced by the activities of mat-building communities of mucilage-secreting microorganisms, mainly filamentous photoautotrophic prokaryotes such as cyanobacteria.

Rather than highlight the link between living and fossil examples, as this definition does, some workers prefer to restrict "stromatolite" to geologic specimens and use "microbial mat" for their living counterparts. Others call structures stromatolites only if they have fairly high relief above the neighboring substrate and refer to the flat-lying ones as mats or sheets. Common to all appropriate definitions is the concept that *it is the biologic origin of the layering in stromatolites that makes them distinctive.* (For this reason, the practice adopted by some of dubbing almost any thinly layered calcareous rock "stromatolitic" should be avoided, because it confuses true stromatolites with nonbiologically deposited look-alikes—cave rocks, such as stalactites and stalagmites, and some hot spring deposits, for instance, formed where minerals build up in thin, sometimes wavy layers as they crystallize from solution.

Fortunately, there are not many kinds of purely inorganically deposited rocks that have stromatolite-like layering, and these usually are fairly easy to tell apart from true stromatolites because they form when minerals repeatedly crystallize out of solution to make stacked surface-coating layers that are almost always more uniform and regular than those laid down by life. And, of course, the purely nonbiologic structures never harbor fossil remnants of mat-building microbial communities.)

As noted in the definition, most stromatolites are calcareous. For this reason, most do not contain structurally identifiable remnants of the microbes that built them (the organic cells having been crushed between growing carbonate grains as the layered sedimentary mass solidified to rock). Hence, like a track, trail, or burrow preserved in an ancient sediment, stromatolites are classed as "trace fossils," organosedimentary structures that evidence biologic activity yet are themselves not fossilized organisms.

On Earth, microbes are so widespread that there is practically no place where stromatolitic look-alikes form without life playing at least a minor role. But on a planet where life never got started there could be many places veneered by thinly layered stromatolite-like deposits unrelated to life—laid down, for instance, by repeated wetting and drying, or freezing and thawing, of mineral-charged salt pans or shallow lagoons. Moreover, it is useful to recall that though fossil stromatolites were recognized as distinctive sedimentary structures early in the 1800s, they were regarded by many geologists as concretionary bodies of purely inorganic origin and it was not until the 1960s (when living specimens were first discovered and fossils of mat-building microbes were found in ancient examples) that their microbial origin was firmly established. Similarly, were stromatolite-like structures detected in an extraterrestrial sample (or photographed on the surface of another planet), it seems certain there would be widespread question as to whether they were produced by life. This uncertainty would be dispelled only if hard and fast rules (based, for example, on diagnostic sedimentological properties) were in place to unambiguously distinguish true stromatolites from nonbiologic mimics, or if the structures were shown to contain cellular remnants of the organisms that built them (in which case, the telling evidence would be the cellular fossils, not the stromatolites themselves).

Cellular Fossils

In studies of ancient life on Earth, optical microscopic studies of two types of preparations have been used to detect remnants of early cellular microbes: acid macerations and petrographic thin sections. Transmission and scanning electron microsocopy have been used to characterize microfossils previously detected in macerations or thin sections, but have not proven to be reliable detection techniques.

Optical Microscopy of Macerations. Maceration, the easier and faster of the two techniques used to detect ancient microfossils, is carried out by dissolving a rock in mineral acid (hydrochloric acid for limestones, hydrofluoric for cherts and siltstones). Because of their coaly composition, organic-walled microfossils pass through the technique unscathed. Abundant fossils are concentrated in the resulting sludgelike acid-resistant residue, which can be slurried onto a microscope slide for study.

Unfortunately, this technique is subject to error-causing contamination. Contaminants can be introduced at almost every stage of the process. At the beginning, even though rock surfaces are cleaned to remove adhering soil, microbes in minute rock crevices are likely to be missed. Laboratory water and commercially available mineral acids used in the technique can also contain a zoo of living contaminants—bacteria, cyanobacteria, unicellular algae, microscopic fungi. And an almost limitless array of fossil-like objects can be introduced during transfer of the residue onto microscope slides. Common culprits include dust, cigarette ash, spores, and pollen grains that settle from the air; lint fibers from clothing or the cloth used to clean microscope slides; small woody fragments and chunks of resin abraded off the wooden rods used in some laboratories to stir the acid-rock sludge; flakes of dandruff and strands of hair; even bits of small spiders that live in water pipes. From the 1950s into the 1970s, as studies of the Precambrian fossil record were getting under way in earnest, all of these maceration-borne contaminants were misinterpreted as fossils by one worker or another (Schopf and Walter, 1983; Mendelson and Schopf, 1992).

Optical Microscopy of Petrographic Thin Sections. In petrographic thin sections, the other type of preparation used for optical microscopic detection of ancient microfossils, fossils are detected *within* the rock, so indigenousness can be demonstrated and the possibility of laboratory contamination ruled out. Consider, for example, microfossils in cherts (together with fine-grained clastic sediments such as siltstones, one of the most fossiliferous rock types known in the early geologic record). Fossil-bearing cherts are made up of cryptocrystalline interlocking grains of quartz laid down from solution. The precipitated grains initially pass through a gem-like opaline state, taking thousands of years to solidify into a full-fledged chert, and the microorganisms are petrified (technically, "permineralized"), embedded within a solid chunk of rock. The quartz grains, which are deposited inside the cells and surround them on all sides, develop so slowly that they grow through the cell walls instead of crushing them. As a result, the petrified fossils are preserved in three-dimensional unflattened bodies that except for their quartz-filled interiors and the brownish color of the aged organic matter that makes them up bear a striking resemblance to living microorganisms. In such sections, only those objects that are entirely entombed in rock can be considered fossil, so it is easy to exclude contaminants that settle onto the surface of a section or are embedded in the resin used to cement the sliver of rock onto the glass thin-section mount.

Optical microscopy of thin sections also provides a way to establish that fossil-like objects date from the time a rock formed rather than having been sealed later in cracks and crevices—that is, to establish that the objects are syngenetic with a primary mineral phase rather than one of secondary or later genesis. Consider, for example, microfossiliferous stromatolitic cherts. When such cherts first form, many contain cavities where gases (often oxygen, carbon dioxide, hydrogen, or methane) given off by the microbial community accumulate in small pockets. Later these cavities can be sealed, filled by a second generation of quartz laid down from seeping groundwater, sometimes tens or hundreds of millions of years after the first chert formed. Microscopic organisms trapped in these cavities and petrified by the second-generation quartz would be true fossils but would be much younger than the rock unit itself. Fortunately, the various generations of quartz in a chert can be sorted out. Rather than having interlocking grains, quartz that fills cavities is a type known as chalcedony that follows the smooth contours of the infilled pocket to form distinctive botryoidal masses. Secondary quartz in cracks or veinlets is also easy to identify because it is angular and its grains much larger than those first formed.

Because special equipment is needed to prepare thin sections—and their study is exceedingly time-consuming—some workers have focused their hunt for ancient fossils on acid-resistant rock residues. In relatively young (Proterozoic) Precambrian rocks, where the fossil record is well enough known that misidentification of contaminants and fossil-like artifacts can be avoided, this technique is useful, simple, and fast. But to avoid mistakes in the oldest (Archean) Precambrian, where the fossil record is not nearly so well known—and, of course, in any extraterrestrial sample—use of the more rigorous thin section technique is essential.

Electron Microscopy as a Detection Technique. Though both transmission electron microsocopy (TEM) and scanning electron microsocopy (SEM) have been used to characterize the cellular makeup and morphology of Precambrian microfossils, neither has proven to be a reliable detection technique.

The TEM studies have involved examinations either of organic-walled fossils freed from their rock matrix by acid maceration, embedded in epoxy resin, and sectioned using a diamond knife, or of fossil-like objects detected in plastic ("formvar") surface replicas of polished and etched petrographic thin sections. Transmission electron microscopy of macerated microfossils has been used to elucidate the structure of cell walls, membranes, and internal organic contents (Schopf and Oehler, 1976). But because macerations are susceptible to contamination, such studies are useful only of fossils previously detected in petrographic thin sections. Similarly, contamination by nonindigenous particles presents a problem for studies of surface replicas, as does the introduction of nonbiogenic artifacts of a variety of types (blisters, bubbles, strands of formvar, and so forth) that in the 1960s and early 1970s were repeatedly identified mistakenly as "ancient fossils" (Schopf and Walter, 1983).

Preparation of samples for study by scanning electron microscopy is simpler than for transmission electron microscopy, and the results obtained are generally easier to interpret. But, as in TEM studies, establishment of the indigenousness of the objects detected and their syngenicity with a primary mineral phase are not straightforward. And, also as in TEM studies, fossil-like artifacts have been misinterpreted by SEM, especially in rock samples where mineralic morphology has been altered and smoothed to "biologic-like" shapes by acid-etching (Schopf and Walter, 1983).

In sum, neither TEM nor SEM has proven reliable for detection of microfossils in the Precambrian rock record on Earth. In view of this track record, claims of detection of minute fossils in extraterrestrial samples by use of electron microscopy should be regarded with skepticism.

Biogenicity. Though optical studies of petrographic thin sections can overcome the problems of establishing indigenousness and syngenicity, the problem of biogenicity often remains. Here, too, lessons learned from studies from the search for Precambrian microbes apply, where in past decades Precambrian microstructures "unlike known mineral forms" have been claimed to be "fossils" simply for want of any other explanation. Living contaminants, "lifelike" dust particles, ball-shaped mineral grains, clumps and shreds of compressed coaly organic matter, solid opaque globules, and a variety of other objects have all been claimed as Precambrian fossils, often on the basis of only one or a few specimens and despite the absence of identifiable cells or other telltale features of living systems (Schopf and Walter, 1983; Mendelson and Schopf, 1992). Many of these reports are founded on the notion that because an object doesn't look mineral it "must" be fossil. This negative reasoning, inference by default, is insufficient. Claims of evidence of past life—whether in Earth rocks or in extraterrestrial samples—need to be backed by positive evidence, hard facts showing what an object actually *is* rather than what it seemingly is not.

Furthermore, the positive evidence adduced must be strong enough to rule out plausible nonbiologic sources. For example, because organic matter can be produced in nonbiologic ways (as when life originated or today in interstellar space), the mere presence of coaly particles in an ancient sediment is not enough to prove that life existed. And because unicell-like organic spheroids can form without life (from clumping of organic matter in seawater or by coaly matter coating ball-shaped mineral grains), tiny round organic bodies in a rock cannot be regarded as assured fossil cells. Nit-picking care of this sort is no longer necessary for reports of fossils from the younger (Proterozoic) part of the Precambrian where evidence of life is overwhelming. But in the older (Archean) part, where so little yet is known—and, even more so, in any extraterrestrial sample—demanding rules must still apply.

Probably the best way to avoid being fooled by nonbiologic structures is to accept as bona fide fossils only those of fairly complex form. This may seem an unreasonably stringent rule for truly ancient fossils since the earliest kinds of cellular life (here and presumably elsewhere) almost certainly were very simple—probably individual, tiny, spheroidal cells. But until we have a sounder base of knowledge and better rules to separate nonfossils from true, it is best to err on the side of caution. For the present, even in ancient terrestrial samples it is safest to accept as biologic only fossils that have unquestionably biologic form, for example colonies of ball-shaped cells embedded in a surrounding organic envelope and thread-like filaments made up of chains of many cells. As evidence builds, we will gain confidence to better interpret less certain finds.

Lessons Learned

On the basis of what has been learned from the search for evidence of Precambrian life on Earth, the biologic origin of putative ancient terrestrial fossils can be accepted if they are (1) made up of coaly organic matter (or are shown to be mineral-replaced); (2) complex enough in *cellular* structure to rule out plausible nonbiologic origins; (3) represented by numerous specimens (if one specimen is preserved, others should be too); and are (4) members of a multicomponent assemblage (terrestrial ecosystems are never monospecific) that (5) exhibit a range of taphonomic degradation consistent with their mode of preservation. In accordance with younger fossils and living microbes, the objects also should be shown to (6) exhibit (gene-based) morphological variability; (7) have inhabited a plausibly livable environment; (8) have grown and reproduced by biologic means of cell division; and (9) exhibit a biogenic isotopic signature.

These nine criteria for biogenicity, and the four other tests that must be met of bona fide terrestrial ancient fossils (provenance, age, indigenousness, and syngenicity) are the product of 30 years of trial, error, and ultimate success. If we seek to avoid Feynman's pitfall of "fooling ourselves," at a minimum we must apply criteria at least as rigorous in the search for evidence of extraterrestrial life, regardless of how minute the putative "microbe-like" objects may be.

References

McKay, D.S., E.K. Gibson, K.L. Thomas-Keprta, H. Vali, C. Romanek, S.J. Clemett, X.D.F. Chiller, C.R. Maechling, and R.N. Zare. 1996. Search for past life on Mars: Possible relic biogenic activity in Martian meteorite ALH84001. *Science* **273**:924-930.

Mendelson, C.V., and J.W. Schopf. 1992. Proterozoic and selected Early Cambrian microfossils and microfossil-like objects. Pp. 865-951 in *The Proterozoic Biosphere, A Multidisciplinary Study,* J.W. Schopf and C.Klein (eds.). New York: Cambridge University Press.

Schopf, J.W. 1993. Microfossils of the Early Archean Apex chert: New evidence of the antiquity of life. *Science* **260**:640-646.

Schopf, J.W., and D.Z. Oehler. 1976. How old are the eukaryotes? *Science* **193**:47-49.

Schopf, J.W., and M.R. Walter. 1983. Archean microfossils: New evidence of ancient microbes. Pp. 214-239 in *Earth's Earliest Biosphere, Its Origin and Evolution,* J.W. Schopf (ed.). Princeton, New Jersey: Princeton University Press.

TAPHONOMIC MODES IN MICROBIAL FOSSILIZATION

Jack Farmer
Department of Geology
Arizona State University

Introduction

The microbial fossil record encompasses a wide variety of information (Figure 1) including morphological fossils (e.g., preserved cellular remains, microfabrics, and stromatolites) and chemofossils (e.g., organic biomarker compounds, isotopic and other geochemical signatures, including biominerals). Observations range in scale from mesoscopic biosedimentary fabrics, to microscopic cellular structures, to sub-microscopic chemical signatures. At all scales of observation, problems often arise when trying to distinguish between biological and inorganic features in the ancient rock record. Stromatolites, defined as laminated biosedimentary fabrics formed by the trapping and binding of sediments and/or precipitation of minerals by microorganisms (Walter 1977), are sometimes impossible to distinguish from finely laminated sediments formed by inorganic processes (see Grotzinger and Rothman 1996). At the cellular level, the biogenicity of Precambrian microfossils has been debated and criteria suggested for identifying pseudofossils (Schopf and Walter 1983; Buick 1984; Awramik et al. 1988). Distinguishing biominerals from their inorganic counterparts has also proven quite difficult in fossil materials, and the biological interpretation of isotopic and organic chemical evidence can also be inconclusive, especially for rocks that have undergone significant diagenesis. The challenge of establishing biogenicity in ancient materials is illustrated by recent debates over the origin of features in Martian meteorite ALH 84001 (McKay et al. 1996; see also review by Treiman 1998). The following discussion is aimed at identifying approaches that may enhance our ability to recognize biosignatures in ancient rocks.

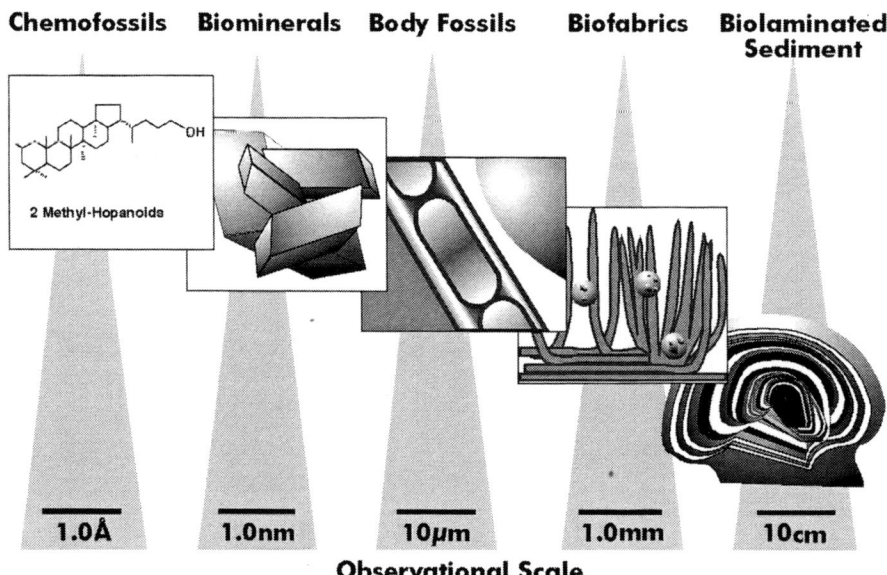

Figure 1. Types of microbial fossil information arranged from right to left in order of decreasing spatial scale.

Importance of Spatially Integrated Studies for Assessing Biogenicity

The most important legacy of ongoing studies of the ALH 84001 meteorite may ultimately prove to be the value of the approach used, namely the integration of evidence over a broad range of observational scales. In developing integrated approaches to the study of biogenicity in ancient materials (whether of terrestrial or extraterrestrial origin), the challenge to the paleontologist is basically threefold. First is to place field samples within a well-defined geological (stratigraphic/age and paleoenvironmental) context. Although the materials may themselves contribute to such knowledge, detailed studies generally demand an understanding of both regional and local geology. Second is the detailed mesoscopic and microscopic characterization of individual samples. Establishing a detailed framework of microscopic observations is requisite for subsequent microsampling of a rock for geochemistry. This involves the identification of mineralogical and microtextural frameworks to distinguish between primary and secondary (diagenetic) features and to establish sequences of mineral paragenesis. At this step it is important to avoid sample contamination or the introduction of structural artifacts during sample preparation that can lead to misinterpretations. The third step involves the application of microsampling methods to place geochemical observations within the microscale spatial and temporal frameworks defined in Step 2. With the development of more spatially refined sampling methods it is now possible to chemically interrogate single mineral phases within rocks. This enables a much more refined understanding of the microenvironmental factors that have affected the preservation of fossil biosignatures.

Even using the spatially integrated approach outlined above, biogenic hypotheses often remain untested. The debate over ALH84001 illustrates this point. Hypothesis testing needs to be more than just confirmatory in nature. Because inorganic processes can so easily confuse biological interpretations in ancient materials, every opportunity should be taken to *disprove* life-based hypotheses. Because we are working with complex historical systems, side-by-side comparisons of modern biological and inorganic analogs may be required to adequately formulate and test hypotheses.

Taphonomic Bias and Common Modes of Preservation

Taphonomy is that subdiscipline of paleontology that deals with the transition of fossil remains from the biosphere to the lithosphere (Efremov 1940; Wilson 1988). In his review of taphonomy, Muller (1979) included a consideration of the cause of death, the processes of decomposition (necrolysis), postmortem transport or other events leading up to burial (biostratinomy) and post-burial events, inclusive of chemical and mechanical changes that occur within the sediment (fossil diagenesis). Taphonomic concepts have been developed largely with reference to multicellular organisms, but the same basic principles apply to the microbial fossil record (e.g., Bartley 1996; Knoll 1985; Knoll and Golubic 1979; Oehler 1976). In the present context, an understanding of common taphonomic modes in microbial fossilization has value in providing additional criteria for assessing biogenicity.

Preservational Modes in Precambrian Marine and Lacustrine Environments

In Precambrian paleontology, studies of organically preserved cellular remains have provided the most complete paleobiological picture of early microbial life. However, the Precambrian record reveals a strong preservational bias based on lithology, paleoenvironment, and the structure of organic materials (Knoll 1985). Perhaps the most instructive Precambrian fossil microbiotas are those preserved as three-dimensional forms in silica or phosphate. Figure 2A provides an example of preservation in silica from the 2.15 Ga. Belcher Island Group of Canada (Hofmann 1976; photomicrograph courtesy of Hans

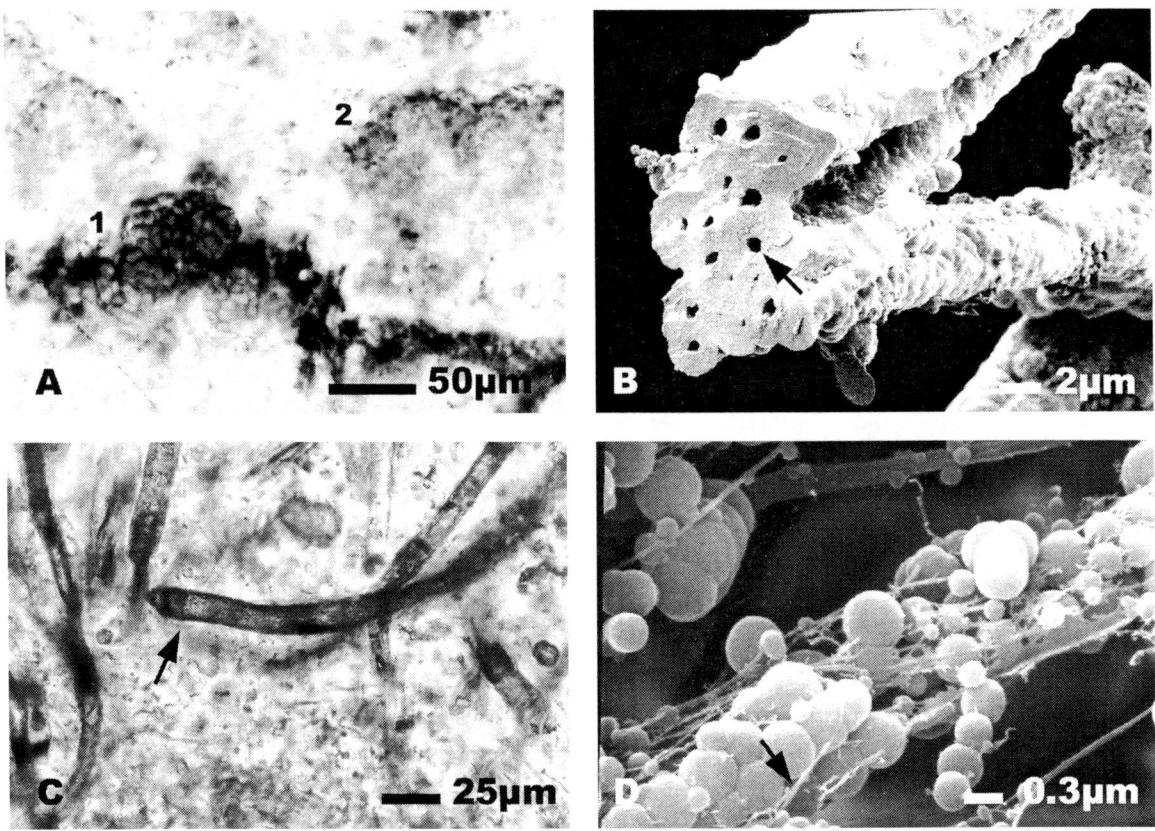

Figure 2. Common taphonomic modes. (A) Thin section photomicrograph of fossilized cyanobacterial mats dominated by a colonial coccoid species of *Eoentophysalis*. Sample from a silicified limestone of the Belcher Island Group, Canada. Early silicification of some laminae enhanced organic matter preservation (compare 1 and 2 on image). Photo courtesy of Dr. Hans J. Hofman. (B) Filament molds of *Phormidium* (filamentous cyanobacteria) formed by encrustation of trichomes. Sample from mid-temperature facies, Excelsior Geyser Basin, Yellowstone National Park. (C) Preserved sheaths of *Calothrix* (filamentous cyanobacteria), within a pisolith, from the Beach Geyser Group, Yellowstone National Park. (D) Critical point dried sample of a *Calothrix* mat showing clusters of silica spheres intimately interspersed within the dried filamentous remnants of an exopolymer matrix (see arrow in C).

Hofmann). In this case, silica was infused into thin mats of *Eoentophysalis,* a colonial coccoid cyanobacterium. In this case, the early silica permineralization of extracellular capsules preserved the external form of individual cells, as well as important aspects of the mat architecture (type of mat growth, stratification of organisms, etc.). Most Precambrian examples of this type were deposited in peritidal marine shorelines of elevated salinity (Knoll 1985). The second major lithotype is fine-grained, clay-rich detrital sediment (shale and volcanic ash) where microbiotas are preserved as two-dimensional compressions flattened during the compaction of sediments. These sediments usually represent deeper basinal settings where anaerobic conditions prevailed. Although several studies have demonstrated that rates of aerobic and anaerobic decay of organic matter do not differ significantly over a broad range of environments (see Lee 1992 and references therein), rates of early diagenetic mineralization have been shown to be higher in anaerobic environments. Such early mineralization is perhaps the most important singular factor in promoting organic matter preservation (Allison 1988; Allison and Briggs 1991).

Taphonomic Trends in Siliceous Thermal Springs

As illustrated by the Belcher Island example, the microbial fossil record is strongly biased toward organisms that possess degradation-resistant cell walls and extracellular envelopes (sheaths and capsules). Another example that illustrates this point is drawn from taphonomic studies of modern and ancient siliceous spring deposits (Farmer and Des Marais 1994; Farmer et al. 1995; Jones and Renaut 1997; Jones et al. 1997; Cady and Farmer 1996; Walter et al. 1998). In rapidly mineralizing thermal springs, the dominant mode of preservation is encrustation of biological surfaces by precipitating minerals, followed by the rapid degradation of organic materials to produce external molds of cells and filaments (Figure 2B). Occasionally, within the lowest temperature facies of thermal spring deposits, partially degraded and permineralized trichomes are observed. But generally speaking, only the sheaths of cyanobacteria (which resist degradation) are preserved in sub-recent fossil materials (Figure 2C).

Comparative taphonomic studies of modern and ancient thermal spring deposits suggest that preservation is strongly skewed toward higher order biofabrics (stromatolites and biologically mediated microtextures) with cellular preservation being limited to lower temperature facies. The fidelity of cellular level preservation appears to increase with decreasing temperature, in large part owing to systematic biological changes that occur along thermal gradients. At temperatures below ~59°C (the upper temperature limit for thermophilic species of *Phormidium,* a *filamentous* cyanobacterium), mats typically increase in thickness, exhibiting a wider variety of surface textures and internal mat fabrics. Below ~35°C, mats similarly show a wide variety of mat structures preserved as stromatolitic fabrics. But in addition, over these temperatures, filamentous cyanobacteria (e.g., *Calothrix*) exhibit substantially larger cell diameters and much thicker sheaths, a factor that enhances cellular level preservation.

Effects of Silica Diagenesis

During the early diagenesis of siliceous spring deposits, there is a pervasive structural reorganization of primary sedimentary fabrics owing to the recrystallization of metastable silica polymorphs (Opal A) to quartz. At the cellular level, only the filamentous and coccoid species having cell diameters larger than ~2 µm and that possess thick extracellular sheaths or capsules, survive diagenetic recrystallization. Thus, within siliceous spring deposits at all scales, the morphological fossil record is heavily biased toward the *filamentous* cyanobacteria (photoautotrophs) and, in particular, larger species that occur below temperatures of ~59°C in modern springs. The smaller coccoid and filamentous Bacteria and

Archaea (cell diameters usually <1 μm) that dominate at temperatures >73°C are also rapidly encrusted by silica. But, in the absence of thick cell walls or extracellular sheaths, entombed organic matter is rapidly decayed away, leaving behind only external molds that are rapidly infilled with opaline silica. During recrystallization there is a tendency for grain size to increase. Fine crystallites of amorphous Opal A (<5 μm; see Figure 2D) are first replaced by microquartz (5-20 μm), which are in turn replaced by megaquartz (20 to >200 μm; see Hesse 1990). This results in the loss of most primary microstructure (see Walter et al. 1998). Because of their small size, the 1-2 μm cell molds of smaller Bacteria and Archaea are quickly obliterated during recrystallization. Acid etching of ancient chert samples sometimes reveals a variety of simple spherical and rod-shaped forms within the matrix of the rock. However, at the microscale, inorganic precipitates of silica exhibit shapes very similar to simple cells, and cell-like forms may also be formed inorganically during the acid etching process. In the absence of cross sectional views that show evidence of a cellular structure, such morphological features are not compelling evidence for biogenicity.

Microbial Biomineralization

An understanding of the varied roles that microorganisms play in mediating mineralization processes is of fundamental importance in understanding fossilization, the origin of biosedimentary fabrics, and the processes of early diagenesis. There are a potentially wide variety of processes that are of interest in this context (see Ehrlich 1996), but for convenience they may be grouped into active processes in which mineral precipitation is directly driven by the metabolic functions of an organism and passive processes in which precipitation is influenced by the structure (e.g., cation-binding) properties of cellular or extracellular materials. In the following discussion, emphasis is placed on passive processes and the importance of extracellular materials in mineralization.

Metal-binding Capacities of Microbial Cell Walls and Extracellular Exopolymers

Natural microbial populations usually secrete large amounts of extracellular exopolymer (EPS), ranging from tightly structured capsules and sheaths around cells to a more loosely bound slime that forms the matrix of microbial biofilms and mats. Decho (1990) identified a wide variety of functions for EPS, including (1) buffering the microenvironment around cells against changes in pH, salinity, and the harmful effects of toxins (e.g., heavy metals); (2) protection against the harmful effects of UV radiation and desiccation during exposure; (3) protection against digestion by grazers; (4) adhesion of biofilms to surfaces; and (5) the concentration of exoenzymes and nutrients required for growth (including dissolved carbon compounds).

Microorganisms adsorb and concentrate many metallic cations required for growth through electrostatic interactions with anionic carboxyl and phosphoryl groups in the cell wall. In addition, however, the exopolymers that surround cells are also very reactive and can readily bind metals such as iron (Konhauser et al. 1993, 1994). EPS is a highly hydrated material (~99% water) possessing an extremely porous fibrillar structure that renders it highly adsorptive. The polysaccharides of EPS possess abundant anionic carboxyl and hydroxyl groups that provide potential binding sites for metals. Of special importance are the carboxyl groups of uronic acids (carboxylated polysaccharides) that correlate strongly with the metal binding capacity of EPS (Kaplan et al. 1987). EPS can bind a wide variety of metals, including Pb, Sr, Zn, Cd, Co, Cu, Mn, Mg, Fe, Ag, and Ni (Decho 1990 and references therein). The metal binding capacity of EPS is strongly influenced by pH, being highest around pH 8 (average seawater). Preliminary studies of mineralization in siliceous thermal springs suggest that much of the silica nucleates within the EPS matrix of mats over a pH range of 8 to 9 (Farmer et al. 1997; see also Figure 2D). Ferris et al. (1989) reported that near neutral pH, microbial biofilms concentrated metals up to 12 orders of magnitude higher than observed under acidic conditions. Neutral to alkaline microenvironments are commonly produced within microbial mats and biofilms through such processes as photosynthesis and sulfate reduction (Krumbein 1979).

Biomineralization as a Factor in Microbial Biosedimentology and Fossilization

Given the ability of EPS to concentrate various metals, it is not surprising that bacteria have been implicated in a wide variety of biomineralization processes. As a result of cellular metabolism, microorganisms alter the chemical microenvironment around the cell, modulating the pH, as well as the concentration of a variety of organic and inorganic solutes. This can induce the large-scale precipitation of authigenic minerals in natural environments. Some examples follow.

Thompson and Ferris (1990) attributed seasonal "whitings" in Green Lake, New York, to the precipitation of calcium carbonate, gypsum, and magnesite by the coccoid cyanobacterium, *Synechococcus*. During photosynthesis, the pH microenvironment around individual cells becomes more alkaline owing to the extraction of CO_2. This can result in supersaturation with respect to the previously mentioned phases. Using TEM, Thompson and Ferris (1990) showed that these minerals actually nucleate on the S-layers (extracellular envelopes) surrounding cells (see Beveridge and Graham 1991 for a discussion of S-layers in bacteria). In Green Lake, *Synechococcus* exhibits a unique double S-layer arrangement. As the outer layer becomes fouled with minerals, it is shed and sinks to the bottom, accumulating on the lake floor as a fine-grained carbonate deposit (micrite). During *Synechococcus* blooms, whole cells can become encrusted by this process and incorporated into the sedimentary record.

Konhauser et al. (1994) found that the bacteria comprising epilithic biofilms of riverine environments sequestered significant amounts of iron, along with smaller amounts of Ca, K, Si, Al, and Mn. Mineralization ranged from Fe-rich EPS capsules to more extensive fine-grained mineral precipitates. Authigenic minerals ranged from complex amorphous (Fe, Al) silicates of variable composition to more silica-rich ordered phases intermediate between glauconite and kaolinite.

Zirenberg and Schiffman (1990) reported the encrustation and replacement of bacterial filaments by metal sulfide minerals and silica in deep sea hydrothermal vent environments. They suggested that bacterially mediated processes may contribute to the formation of base-metal sulfide deposits by concentrating silver, arsenic, and copper from seafloor hydrothermal fluids. At lower temperatures, pyrite (FeS_2) is a common mineral phase in fine-grained, organic-rich marine sediments. It is formed under anaerobic conditions by the reaction of iron-bearing detrital minerals in sediments and H_2S produced by bacterial sulfate reduction (Canfield and Raiswell 1991).

Some of the most spectacular examples of microbial fossilization involve the permineralization of organic materials by phosphate minerals (e.g., Xiao et al. 1998). Piper and Codespoti (1975) suggested that the precipitation of carbonate fluorapatite (phosphate) in marine environments may be controlled by the bacterial denitrification of anoxic sediments at sites where the oxygen minimum zone intersects the seafloor. The loss of nitrogen results in a decline in microbial production and the secretion of bacterial phosphatases (see Ehrlich 1996). Such processes may govern the precipitation of phosphate minerals in seafloor sediments, thus favoring the early diagenetic mineralization and fossilization of organic materials (e.g., Rao and Nair 1988).

Discussion

During this workshop both theoretical and empirical approaches have converged on a minimum cell diameter of between 200-300 nm for free-living organisms. However, taphonomic (preservational) biases place different constraints on the lower size limit for *fossil* microbes. Comparative taphonomic studies of the Precambrian fossil record and modern analogs indicate a strong preservational bias, favoring higher order biosedimentary structures and biofabrics. In providing evidence of biogenicity, such features are often not definitive. In contrast, while providing more compelling evidence for biogenicity, organically preserved cellular structures are also much rarer in the record. In addition, taphonomic biases favor microorganisms that are larger than a few microns in diameter and that possess thick cell walls and/or extracellular structures. This constitutes a strong taphonomic filter that excludes most smaller organisms from entering the record.

In seeking an answer to the question posed to this panel—Can we understand the processes of fossilization and inorganic chemistry sufficiently well to differentiate fossils from the artifacts in a sample?—the preceding examples suggest potential directions for future study. Microorganisms mediate a wide variety of natural mineralization processes. To a large extent this appears to be underpinned by the seemingly universal adaptive value of extracellular exopolymers in regulating the biology of microorganisms. The strong tendency of EPS to scavenge a wide variety of metallic cations from the surrounding environment suggests that we may improve our ability to detect biogenic signatures in rocks by searching well-characterized samples for anomalous concentrations of trace metals. In conjunction with other types of chemofossil evidence (e.g., isotopes and organic biomarker compounds), spatial distributions of trace metals that are comparable in pattern and scale to microbial cells and biofilms may provide additional evidence for biogenicity. And through an improved understanding of the varied role(s) played by trace elements in modern microbial processes, we may eventually be able to extract paleobiological information from rocks even where primary organic materials have been completely degraded and lost. The limiting factor is likely to be the survival of trace element biosignatures during diagenesis, a problem that can be addressed through detailed comparisons of modern and ancient analogs.

References

Allison, P.A. 1988. The role of anoxia in the decay and mineralization of proteinaceous macro-fossils. *Paleobiology* **14**: 139-154.

Allison, P.A., and D.E.G. Briggs. 1991. Taphonomy of non-mineralized tissues, Pp. 25-70 in *Taphonomy: Releasing the Data of the Fossil Record*. P.A. Allison and D.E.G. Briggs (eds.). New York: Plenum Press.

Awramik, S.M., J.W. Schopf, and M.R. Walter. 1988. Carbonaceous filaments from North Pole, Western Australia: Are there fossil bacteria in Archean stromatolites? A discussion. *Precambrian Research* **39**: 303-309.

Bartley, J.K. 1996. Actualistic taphonomy of cyanobacteria: Implications for the Precambrian fossil record. *Palaios* **11**: 571-586.

Beveridge, T.J., and L.L. Graham. 1991. Surface layers of bacteria. *Microbiological Reviews* **55**: 684-705.

Buick, R. 1984. Carbonaceous filaments from North Pole, Western Australia: Are there fossil bacteria in Archean stromatolites? *Precambrian Research* **24**: 157-172.

Cady, S.L., and J.D. Farmer. 1996. Fossilization processes in siliceous thermal springs: Trends in preservation along thermal gradients. Pp. 150-173 in *Evolution of Hydrothermal Ecosystems on Earth (and Mars?)*. G. Bock and J. Goode (eds.). Chichester: John Wiley & Sons Ltd.

Canfield, D.E. and R. Raiswell. 1991. Pyrite formation and fossil preservation. Pp. 337-387 in *Taphonomy: Releasing the Data of the Fossil Record*. P.A. Allison and D.E.G. Briggs (eds.). New York: Plenum Press.

Decho, A.W. 1990. Microbial exopolymer secretions in ocean environments: Their role(s) in food webs and marine processes. *Oceanography and Marine Biology Annual Reviews* **28**: 73-153.

Efremov, J.A. 1940. Taphonomy; a new branch of geology. *Pan-American Geologist* **74**: 81-93.

Ehrlich, H.L. 1996. *Geomicrobiology.* (Second Edition). New York: Marcel Dekker Inc.

Farmer, J., B. Bebout, and L. Jahnke. 1997. Fossilization of coniform *(Phormidium)* stromatolites in siliceous thermal springs, Yellowstone National Park. *Geological Society of America, Abstracts with Program* **29**(6): 113.

Farmer, J.D., and D.J. Des Marais. 1994. Biological versus inorganic processes in stromatolite morphogenesis: Observations from mineralizing systems. Pp. 61-68 in *Microbial Mats: Structure, Development and Environmental Significance.* NATO ASI Series in Ecological Sciences. L.J. Stal and P. Caumette (eds.). Springer Verlag.

Farmer, J.D., S.A. Cady, and D.J. Des Marais. 1995. Fossilization processes in thermal springs. *Geological Society of America, Abstracts with Programs* **27**: 305.

Ferris, F.G., S. Schultze, T. Witten, W.S. Fyfe, and T.J. Beveridge. 1989. Metal interactions with microbial biofilms in acidic and neutral pH environments. *Applied and Environmental Microbiology* **55**: 1249-1257.

Grotzinger, J.P., and D.H. Rothman. 1996. An abiotic model for stromatolite morphogenesis. *Nature* **383**: 423-425.

Hesse, R. 1990. Silica diagenesis: Origin of inorganic and replacement cherts. *Geological Association of Canada, Geoscience Canada Reprint Series* **4**: 253-275.

Hofmann, H.J. 1976. Precambrian microflora, Belcher Islands, Canada: Significance and systematics. *Journal of Paleontology* **50**: 1040-1073.

Jones, B., and R.W. Renaut. 1997. Formation of silica oncoids around geysers and hot springs at El Tatio, northern Chile. *Sedimentology* **44**: 287-384.

Jones, B., R.W. Renaut, and M.R. Rosen. 1997. Biogenicity of silica precipitation around geysers and hot-spring vents, North Island, New Zealand. *Journal of Sedimentary Research* **67**: 88-104.

Kaplan, D., D. Christiaen, and S. Arad. 1987. Chelating properties of extracellular polysaccharide from *Chlorella* spp. *Applied and Environmental Microbiology* **53**: 2953-2956.

Knoll, A.H. 1985. Exceptional preservation of photosynthetic organisms in silicified carbonates and silicified peats. *Philosophical Transactions of the Royal Society of London*, Part B, **311**: 111-122.

Knoll, A.H., and S. Golubic. 1979. Anatomy and taphonomy of a Precambrian algal stromatolite. *Precambrian Research* **10**: 115-151.

Konhauser, K.O., S. Schultze-Lam, F.G. Ferris, W.S. Fyfe, F.J. Longstaffe, and T.J. Beveridge. 1994. Mineral precipitation by epilithic biofims in the Speed River, Ontario, Canada. *Applied and Environmental Microbiology* **60**: 549-553.

Konhauser, K.O., W.S. Fyfe, F.G. Ferris, and T.J. Beveridge. 1993. Metal sorption and mineral precipitation by bacteria in two Amazonian river systems: Rio Solimoes and Rio Negro, Brazil. *Geology* **21**: 1103-1106.

Krumbein, W.E. 1979. Calcification by bacteria and algae. *Biogeochemical Cycling of Mineral-Forming Elements.* P.A. Trudinger and D.J. Swaine (eds.). New York: Elsevier Science.

Lee, C. 1992. Controls on organic carbon preservation: The use of stratified water bodies to compare intrinsic rates of decomposition in oxic and anoxic systems. *Geochemica et Cosmochemica Acta* **56**: 3323-3335.

McKay, D.S., E.K. Gibson, K.L. Thomas-Keprta, H. Vali, C.S. Romanek, S.J. Clemett, X.D.F. Chillier, C.R. Maechling, and R.N. Zare. 1996. Search for past life on Mars: Possible relic biogenic activity in Martian meteorite ALH 84001. *Science* **273**: 924-930.

Muller, A.H. 1979. Fossilization (taphonomy). Pp. A1-A78 in *Treatise on Invertebrate Paleontology, Part A: Introduction.* R.A. Robison and C. Teichert (eds). Boulder, Colorado: Geological Society of America and University of Kansas.

Oehler, J.H. 1976. Experimental studies in Precambrian paleontology: Structural and chemical changes in blue-green algae during simulated fossilization in synthetic chert. *Geological Society of America Bulletin* **87**: 117-129.

Piper, D.Z., and L.A. Codespoti. 1975. Marine phosphorite deposits in the nitrogen cycle. *Science* **179**: 564-565.

Rao, P.V., and R.R. Nair. 1988. Microbial origin of the phosphorites of the western continental shelf of India. *Marine Geology* **84**: 105-110.

Schopf, J.W., and M.R. Walter. 1983. Archean microfossils: New evidence of ancient microbes. Pp. 214-239 in *Earth's Earliest Biosphere, Its Origin and Evolution.* J.W. Schopf and C. Klein (eds.). Princeton, N.J.: Princeton University Press.

Thompson, J.B., and F.G. Ferris. 1990. Cyanobacterial precipitation of gypsum, calcite and magnesite from natural alkaline lake water. *Geology* **18**: 995-998.

Treiman, A.H. 1998. The history of Allan Hills 84001 revised: Multiple shock events. *Meteoritics & Planetary Science* **33**: 753-764.

Walter, M.R. 1977. Interpreting stromatolites. *American Scientist* **65**: 563-571.

Walter, M.R., S. McLoughlin, A.N. Drinnan, and J.D. Farmer. 1998. Paleontology of Devonian thermal spring deposits, Drummond Basin, Australia. *Alcheringa* **22**: 285-314.

Wilson, M.V.H. 1988. Paleoscene #9. Taphonomic processes: Information loss and information gain. *Geoscience Canada* **15**: 131-148.

Xiao, S.H., Y. Zhang, and A.H. Knoll. 1998. Three-dimensional preservation of algae and animal embryos in a Neoproterozoic phosphorite. *Nature* **391**: 553-558.

Zirenberg, R.A., and P. Schiffman. 1990. Microbial control of silver mineralization at a seafloor hydrothermal site on the northern Gorda Ridge. *Nature* **348**: 155-157.

INVESTIGATION OF BIOMINERALIZATION AT THE NANOMETER SCALE BY USING ELECTRON MICROSCOPY

John Bradley
MVA Inc. and School of Materials Science and Engineering
Georgia Institute of Technology

Introduction

The search for microbial life in terrestrial and extraterrestrial rocks has recently intensified following the announcement of evidence of past Martian life in a meteorite from Mars [1]. Although there is debate about whether a compelling case has been made for evidence of past Martian life in the meteorite, there is no debate that the evidence exists at the nanometer scale [2,3]. Biomarkers include both organic and inorganic species, although inorganic "biominerals" are perhaps more likely to survive geological processing. Microorganisms that precipitate biominerals during their life cycles can exert control over crystal size, crystallographic orientation, degree of crystal perfection, and morphology. In principle, specific biominerals (e.g., magnetite and Fe sulfides) may be used as indicators of past biogenic activity, providing their properties are significantly different from minerals produced by non-biological processes. Evidence of biomineralization may exist only at the nanometer scale [3]. (Biominerals ~10 nm in diameter and containing less than 10,000 atoms have been observed.) One of the biggest challenges in looking for evidence of past (or present) microbial life in geological samples is to develop and refine analytical methods to probe specimens on a scale comparable to that of the biogenic activity.

Electron Microscopy

Electron microscopy is unique among analytical techniques in that it provides the ability to examine the morphologies, internal structures, crystallography, and compositions of materials at close to atomic resolution. The essential elements of an electron microscope are a high-vacuum column, an electron gun (a thermal or field emission electron emitter), a system of magnetic lenses to focus the electrons before (and after) interacting with the specimen, beam-scanning coils for rastering the electron beam across the specimen. A variety of (electron and x-ray) detectors are available for imaging and spectroscopy.

The two major classes of electron beam instruments are the scanning electron microscope (SEM) and the transmission electron microscope (TEM). (Both instruments have proven useful for studying microorganisms and biominerals). Each instrument exploits a specific electron optical configuration and incident beam energy range that targets it toward certain types of microanalysis. The SEM is used for characterizing the surfaces of thick (electron opaque) specimens (Figure 1). Most SEMs operate in the 2-30 keV range and are configured primarily for imaging (using secondary and backscattered electrons) and compositional analysis (using energy-dispersive x-ray spectroscopy (EDS)). Some SEMs are also equipped with one or more crystal spectrometers for compositional analysis using wavelength-dispersive x-ray spectroscopy (WDS). (WDS offers ~10X better detection limits over EDS for some elements.)

Sample preparation can be of critical importance in SEM. If a specimen is a good conductor, secondary electron images of surfaces with nanometer-scale resolution are possible. If a specimen is a poor conductor or insulator, a conductive coating must first be applied in order to obtain the highest resolution images (see Figure 1). Thin (1-20 nm thick) coatings of carbon, chromium, palladium, or

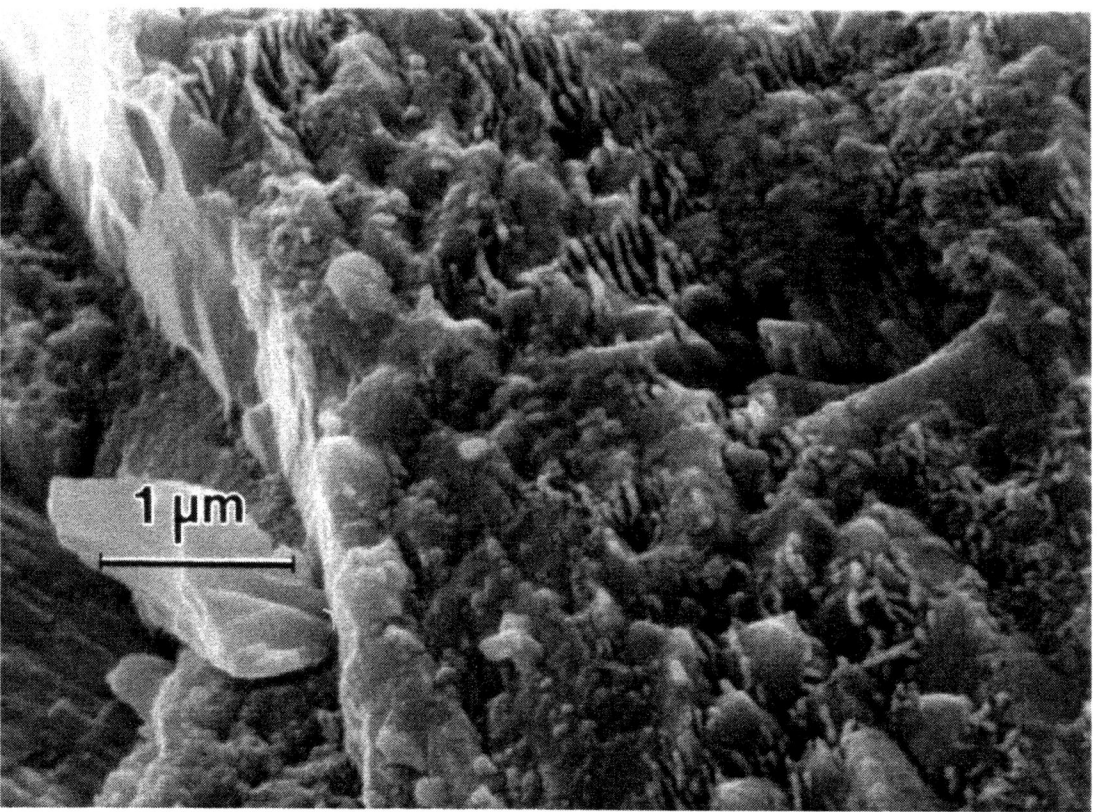

Figure 1. Worm-like elongated forms on a fracture surface within the martian meteorite ALH84001. Since the orientations of many of the elongated forms are parallel to the substrate cleavage direction (vertical ledge at left), it is highly likely that they are mineral lamellae (rather than "nanofossils") with segmented surface structures resulting from deposition of a conductive gold coating [4].

gold are evaporated or sputtered onto specimens to make them conductive. The less conductive the specimen, the more coating must be applied to obtain highest-quality images. However, once a coating has been applied it is primarily the coating rather than the underlying specimen that is being imaged. When imaging nanometer-sized features on a coated surface, great care must be taken to distinguish indigenous surface microstructures from those caused or accentuated by application of the conductive coating. The problem of conductive coating artifacts is particularly problematical with the new generation of field emission scanning electron microscope (FE SEM), because the subnanometer field emission electron beam permits secondary electron imaging with resolution of 1-2 nm. Under these circumstances, coating microstructures that are not resolvable using a lower-resolution SEM are easily resolved using FE SEM.

TEM is used primarily for examination of the interiors of thin (electron transparent) specimens. Most TEMs operate in the 100-400 keV range. A TEM without beam-scanning capabilities is referred to as a conventional TEM, or CTEM, and a TEM equipped with beam-scanning coils is called a scanning TEM, or STEM. Modern analytical STEMs equipped with secondary and backscattered electron detectors provide most of the capabilities of an analytical SEM plus an additional range of capabilities that are specific to the TEM. These include brightfield and darkfield imaging, high-resolution lattice-fringe imaging (Figure 2), electron diffraction, and electron energy-loss spectroscopy. A STEM with a field emission electron gun (FE STEM) offers high beam currents in extremely small electron "nanoprobes" (0.5-1 nm diameter). Coupled with high collection efficiency solid state x-ray detectors, this makes quantitative compositional EDS microanalysis and compositional mapping with spatial resolution of a few nanometers possible. Using the newly emerging electron energy-loss (energy-filtered) imaging

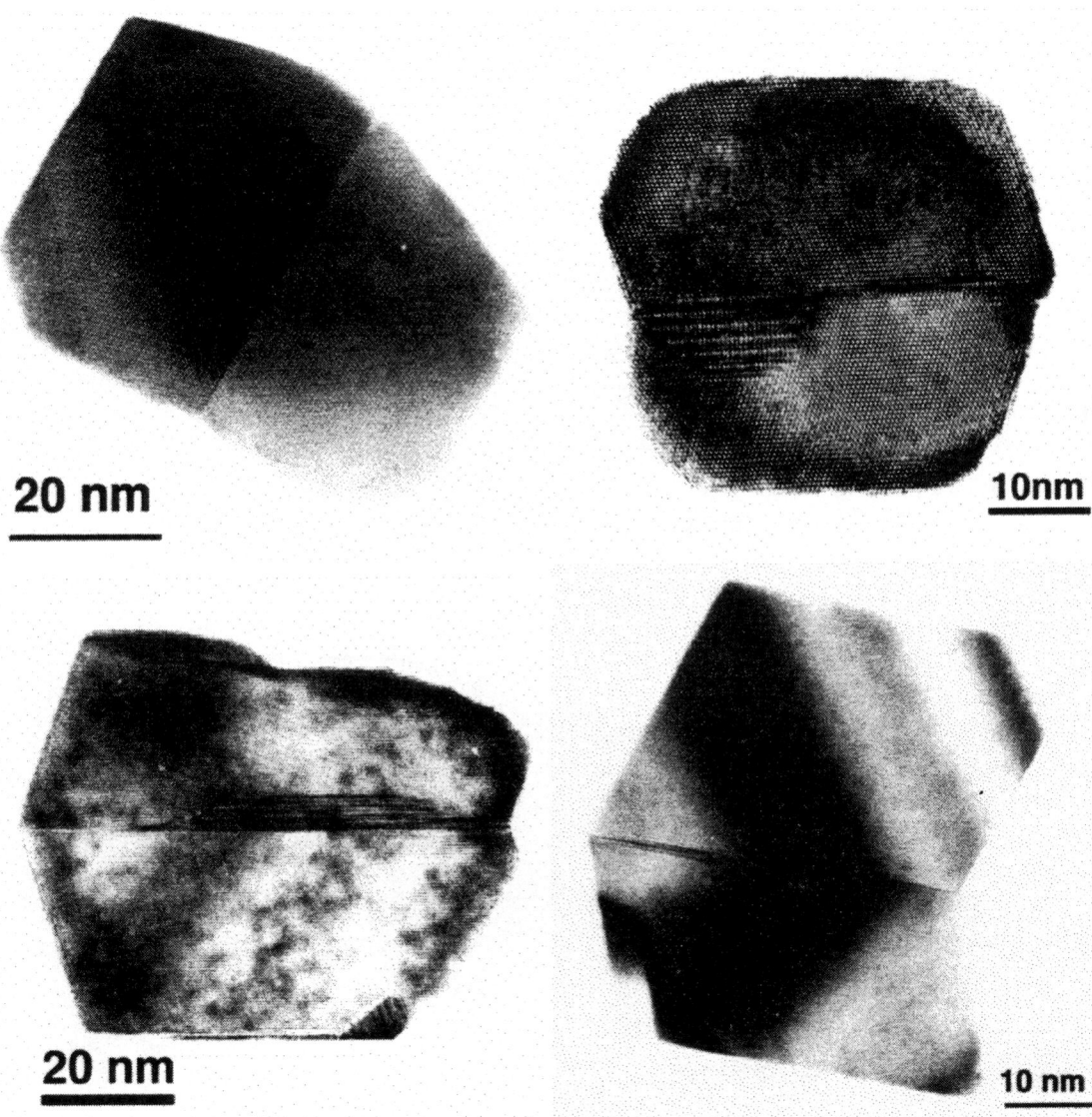

Figure 2. Comparison of twinned biogenic and non-biogenic single-domain magnetite (Fe_3O_4) nanocrystals. The upper-left and lower-right TEM lattice-fringe images are of synthetic magnetite (Fe_3O_4), while the other two are bacterial magnetosomes. The morphological, structural, and crystallographic properties of the biogenic magnetites overlap those of the inorganic magnetites. (Images courtesy of M. Pósfai and P.R. Buseck.)

technology, compositional mapping with resolution ~ 1 nm is possible. Unlike EDS mapping, energy-filtered imaging is only semi-quantitative, but it offers the huge advantage of especially high (collection) efficiency for light element analysis and mapping. Thus, biogenic nanostructures containing organic and inorganic matter could be mapped using energy-filtered imaging. Because thin specimens (ideally <100 nm thick) are required for TEM, specialized sample preparation procedures are required. Ultramicrotomy, ion milling, chemical etching, and precision polishing are the most commonly used methods for producing thin TEM specimens.

Electron Microscopy of Biominerals

Three approaches are potentially useful for detecting evidence of biomineralization in rocks using electron microscopy. They are *morphological* studies using high-resolution SEM imaging, *mineralogical* studies using TEM, and *compositional* studies using TEM.

The morphological approach usually relies on using high-resolution SEM imaging to identify shapes or forms on surfaces (e.g., worms) that are consistent with past biological activity. This approach has been used to search for nanofossils in meteorites and terrestrial rocks [1,5]. However, image interpretation is subject to uncertainties, and it is usually difficult to obtain corroborating compositional and structural data from the same specimen [4]. Conductive coatings produce nanometer-sized morphological forms that have been confused with biological forms [5]. A variety of exotic morphological forms similar to biogenic structures can be produced by strictly non-biological processes [4,6]. Even if the morphology of a particular form is consistent with biogenic activity, it may not be unique to biogenic activity.

The TEM has proven ideal for probing the mineralogy of biominerals [2,3]. Common biominerals include iron oxides (e.g., magnetite) (see Figure 2), iron sulfides (e.g., greigite and pyrrhotite), carbonates, and other minerals. Some biominerals are arranged in distinctive configurations. For example, magnetotactic bacteria are a group of organisms that orient and navigate along geomagnetic field lines, and they do so by precipitating chains of magnetite (or iron sulfide) nanocrystals. Unfortunately, the chains may not survive geological processing, and the individual bacterial magnetosomes that make up the chains can be difficult to distinguish from some inorganically produced magnetites (see Figure 2).

Compositional analyses at the nanometer scale can be useful for investigating biogenic structures. The distribution of heavy elements can be mapped with resolution on the order of ~5 nm using EDS. Electron energy-loss energy-filtered imaging can be used to investigate the distribution and speciation of biogenically important light elements C, N, and O at the nanometer scale. Organic compounds (e.g., PAHs) associated with potential biominerals may be indicators of past biogenic activity [1]. Although molecular species cannot be directly detected using electron microscopy, it is possible to probe the local (atomic and molecular) bonding environment of C, N, and O (and heavier elements), using electron energy-loss spectroscopy.

References

1. D.S. McKay et al. (1996). *Science* **273**, 924-930.
2. M. Pósfai et al. (1998). *Science* **280**, 880-883.
3. A. Iida and J. Akai (1996). *Sci. Rep. Niigata Univ., Ser E (Geology)* **11**, 43-66.
4. J.P. Bradley et al. (1997). *Nature* **390**, 5145-5146.
5. V.A. Pedone and R.L. Folk (1996). *Geology* **24**, 763-765.
6. R. Symonds (1993). *Geochem. J.* **26**, 337-350.

Panel 4

Does our current understanding of the processes that led from chemical to biological evolution place constraints on the size of early organisms?

If size is not constrained, are there chemical signatures that might record the transition to living systems?

DISCUSSION

Summarized by Leslie Orgel, Panel Moderator, and Laura Ost, Consultant

Did Life Originate in an RNA World?

Free-living organisms today require two biopolymers—DNA and RNA, which store and transmit genetic information—as well as proteins, which catalyze chemical reactions. A primitive organism might have relied on a single biopolymer, RNA, which might or might not have catalyzed its own reactions. Such an organism would not have required proteins, ribosomes, and other modern cellular machinery and conceivably could have been the smallest self-sustaining chemical system capable of Darwinian evolution.

If life began with RNA, then it must have started with nucleotides made accidentally in a prebiotic process, but it must also have been "clever enough" to invent the materials needed for the next phase of evolution, Dr. Orgel observed. Panel members offered varying perspectives on how this may have occurred, reflecting not outright disagreements but rather different areas of expertise and interest and perhaps different phases of evolution.

Dr. Ferris suggested that, instead of trying to "downsize a Mercedes into a Yugo," it might be useful to think in terms of the comic strip B.C.'s round stone with a stick through the center. He proposed that RNA-based life-forms originated from monomers present on the primitive Earth. RNA may have catalyzed its own reactions, and other necessary compounds might have been formed in a variety of ways. Bases formed from hydrogen cyanide in aqueous solution could produce adenine and guanine, and purines could be generated from these two compounds or, alternatively, brought in by meteorites. Formaldehyde can be converted to ribose and many other sugars. Montmorillonite clay acts as a catalyst in making RNA oligomers, which, once they are long enough (perhaps 50 mers), may have catalytic properties.

In Dr. Ferris's concept, the RNA, vesicles, and proteins—not enclosed by a cell membrane—would bind to mineral surfaces. Their shape and dimensions would be determined by the features of the substrate and rates of formation of RNA. These organisms could be as small as some purported nanobacteria, or about twice the size of the Q-beta virus, which contains three genes consisting of about 1,500 bases each. Dr. Ferris concluded that the fossil signature of such RNA-based life-forms would be difficult to identify.

Dr. Szostak proposed that a simple ancestral cell with the capability to evolve into a more complex cell may have started with polynucleotides, which can have catalytic activity, and vesicles, which are spontaneously assembling systems. Once a replicase and vesicle are brought together, a synergistic evolution could build up to produce a megabase of information that leads to a free-living organism. This process sets the stage for peptide synthesis and large-scale structural and regulatory components.

Dr. Szostak observed that evolution is inhibited by the free interaction of replicases in solution. The only way to achieve interesting Darwinian evolution is to have a compartmentalized system that can grow and divide, thus providing a selective advantage for mutations. But how can there be a cell cycle without any internal encoded machinery? Small vesicles, 30 to 100 nm in size, could interact and fuse to generate larger ones that combined different internal molecules. In the laboratory an artificial system can be created in which cells divide, fuse, divide, and fuse. Much larger vesicles can be fragmented with mild shear forces.

Dr. Benner proposed that the minimum cell size would be determined by the robustness of a single-biopolymer system in making the chemical compromise between genetics and catalysis, which pose competing and contradictory demands (e.g., in terms of the biopolymer's complexity, ease of folding, and capability to change physical properties). The problem is that nucleic acids are generally not good catalysts: one must sort through 2×10^{13} random RNA sequences to find one that modestly increases the rate of a templated ligation. Adding functional groups does improve catalytic power and versatility, but it is not clear whether functionalized RNA can sustain Darwinian evolution.

Dr. Benner said that chemical studies attempting to resolve these contradictions will help define life's origins on Earth and how best to find life elsewhere. In the meantime, short of historical context, information content is the only reliable signature of a Darwinian chemical system. A single-biopolymer system must be able not only to replicate but also to evolve. There is evidence that life before proteins had functionalized RNA, so this chemistry should be sought in samples from Mars. He also proposed that a genetic molecule needs a polycharged backbone to exhibit the behavior needed to support Darwinian evolution. Such a chemical structure would be fairly easy to detect on Mars, perhaps robotically.

The First Biopolymer System

A question was raised concerning how the first biopolymer was formed, given that even modern cells must work hard to make highly activated molecules. In fact, as Dr. Orgel noted, this is a matter of dispute within the prebiotic research community. Dr. Ferris said that a primitive process for forming such molecules is plausible, because polymeric phosphates can be made by heating phosphates.

Dr. Fraenkel asked whether prebiotic evolution would have been assisted by high temperatures. Dr. Benner noted that high temperature speeds all reactions, whether desirable or not, and it destroys the secondary structure of nucleic acids. Dr. de Duve noted that some scientists believe that life originated at cold temperatures—below zero degrees centigrade.

Dr. Ferris said that scientists have been looking for a polymer system other than RNA that could have driven early life-forms, but they have failed so far, so RNA remains the best model. Dr. Orgel noted that some other systems behave much like RNA and DNA but have modified backbones. Peptide nucleic acid, for example, which lacks the components that form the deoxyribosyl-phosphate backbone, may be marginally simpler than RNA. However, it has been difficult to synthesize.

The First Cell Membrane

The issue of encapsulation was revisited by Dr. Orgel, who asked when, on the evolutionary scale, an impermeable cell membrane stopped being a disadvantage and became a necessity. A cell that depends on external resources produced by abiotic processes clearly cannot obtain them if it is surrounded by such a membrane. Similarly, a cell that makes its own metabolites cannot allow them to escape. Dr. de Duve suggested that when encapsulation evolved, it enabled competition between cells instead of molecules.

Dr. Szostak said that permeable membranes might have been formed from short-chain fatty acids or alcohols. A cell with an impermeable membrane would need to be complex enough to both evolve the barrier and encode a transport system, perhaps with nucleic acids or peptides serving transport functions. Dr. Benner said that membranes likely to emerge in a primitive environment would contain multiple organic molecules, would be defective, and would be permeable. He suggested phosphorylation as a way of holding resources inside leaky membranes. Dr. Ferris said that nutrient flow might be restricted and noted that no peptides would have been available in the RNA world. Dr. Osborn postulated that inorganic phosphates inside and outside the cell might reach an equilibrium through a leaky membrane. But she asked whether any intrinsically leaky membranes are known; membranes are not made of fatty acids, but rather from phospholipids.

If a membrane-like structure is observed in a sample, then how big must it be before it can realistically be considered a cell membrane, and does the cell need multiple genes, Dr. Orgel asked. Dr. Szostak suggested that tens or hundreds of genes would be needed; a one-gene cell could not encode transport and would have a leaky membrane. Dr. Fraenkel asked what types of molecules would need to be transported—nucleotide triphosphates? Resources such as carbon dioxide can pass through modern membranes without a transport system, but phosphates cannot. Dr. Benner said that his comments refer to life-forms that are just sophisticated enough to achieve a metastable state—probably the type most likely to be found on Mars. By contrast, Dr. Szostak said that there is no reason to think that the evolution of protein synthesis is difficult.

Time Frame for Evolution of Life on Mars

Earth was formed 4.5 billion years ago, and approximately 1 billion years later bacteria resembling modern cyanobacteria had evolved. When, and for how long, did Mars offer a suitable environment (i.e., water) for evolution? Just as scientists are uncertain about what it takes to create a fully competent organism, so also is the time frame for the aqueous history of Mars subject to debate, although the surface had probably dried out about 3 billion years ago.

Dr. Ferris said that life is unlikely to have survived on the surface of Mars because of the inhospitable environment. Presumably life originated and was shut down quickly. John Rummel said that life may have evolved over a long time period because the massive outflow features on Mars suggest that there may be large quantities of liquid water beneath the permafrost. There may also be subsurface volcanic or hydrothermal activity. Furthermore, as part of the natural cross-contamination between celestial bodies, terrestrial materials bearing viable organisms may have been transported to Mars at a time when water flowed on the surface.

Dr. Benner said that the search for life on the martian surface is a surrogate for the search for life elsewhere. He assumed that life would have emerged on Mars and on Earth at about the same time. One billion years is 10 percent of the life of a star; of the billions and billions of planets, scientists are examining the one planet most readily available. If it took a long time for protein translation to emerge, then martian life-forms might exhibit the primitive character of a single-biopolymer system, but they might have made the transition to a two-polymer system.

Dr. de Duve said it is unlikely that life took a long time to emerge because it involved chemical reactions. He proposed that life arose rapidly, perhaps many times, until it finally was sustained. He disputed the notion of tiny protocells harboring RNA molecules that exhibited both genetic and catalytic activity swimming in a sea of activated nucleotides. The nucleotides would not feed through the cell membrane to enable adenosine triphosphate or guanosine triphosphate to pass through. Although the basic premise of the RNA world may be correct, he said that a complex proto-metabolism was needed that was catalyzed by clays, metals, or peptides instead of RNA molecules, because a catalyst was needed to make the first RNA molecule.

Dr. Szostak agreed that primitive organisms might have evolved quickly, adding that it might have taken just a few years to evolve from a one-gene cell to a free-living organism with perhaps 100 genes.

Minimum Cell Size

Dr. Orgel said that the discussion suggested that a replicating, single-biopolymer system could be compressed into a very small volume just slightly larger than the genome, in contrast to two-biopolymer systems, which must be at least 5 to 10 times the volume of the genome. Dr. Benner suggested that a single-biopolymer system could be packed into a 50-nm sphere. Thus, although a sphere of 50 nm in a terrestrial sample would not represent a life-form, a similar structure in a martian sample would warrant study of the organic chemistry to determine whether it had a genome.

Summary and Consensus

As yet, there is no consensus view of how life originated. There is, however, broad agreement that the first living systems were far simpler than the simplest free-living organisms known today. The concept that life passed through a stage in which RNA, or a polymer much like it, provided both genetic information and catalysis suggests what such a simple organism might have been like. Organisms characterized by such single-bioploymer chemistry could have been minute, perhaps as small as 50 nm in diameter. This means that the minimum size observable in living cells may not be applicable in setting limits for biological detection on Mars or Europa. The earliest organisms on Earth (or elsewhere) would probably be extremely difficult to recognize as fossils.

PRIMITIVE LIFE: ORIGIN, SIZE, AND SIGNATURE

James P. Ferris
Department of Chemistry
Rensselaer Polytechnic Institute

Abstract

The question of the size of the first life was brought to the forefront by the proposal that the martian meteorite ALH84001 contains nanometer size fossils of martian life. In this paper estimates of the size of the first life were made on the basis of the essential requirements for life and research progress toward the understanding of the origin of life. One model for the first life is based on RNA bound to the mineral that catalyzed the formation of the RNA. The essential life processes, with the exception of the synthesis of monomers, were carried out by the RNA bound to the mineral surface. The size of this life was determined by the size of the mineral surface and the rates of formation and decomposition of the RNA. The second model for the first life assumes that the RNA and other essential biomolecules were protected from dispersal by a membrane. Here it is assumed that synthesis of monomers took place within the membrane. The size of this life was estimated from the sizes of RNA viruses, and it was concluded that the first life could have been as small as the proposed "nannobacteria."

Introduction

The proposal that the Mars meteorite ALH84001 contained fossils of "nannobacteria" (McKay et al., 1996) prompted, among other discussions, one on the minimal size for life. A point in favor of such small life-forms is that the first life on Earth and Mars would have been much smaller and simpler than the present life on Earth, so comparisons to contemporary cellular life are probably not valid. I will examine the possible size and shape of the first life on Earth and/or Mars. This will be done by reviewing some of the experimental data concerning the pathway to the first life. I will then extrapolate from that data to two different possibilities for the first life and then use these models for life together with the known sizes of the genomes of viruses to estimate the sizes of the first life. Finally, the possibilities of finding fossil signatures of this life on Earth and Mars will be discussed.

What Is Life?

It will be necessary to provide a definition of the basic requirements of life before it is possible to suggest what constitutes a minimal form of life. What is life? is a controversial scientific question because it is intimately associated with the particular scenario that the scientist is investigating for the origin of life. He/she does not want a definition that would invalidate their paradigm of the origin of life. The definition of life was the topic of a recent paper by Luisi (1998). He provided a brief review of the definitions put forward over the past 100 years and then focused his discussion on recent definitions. The simplest is, "Life is a self-sustaining chemical system undergoing Darwinian evolution." He proposed a modification of this definition for "adherents of the RNA world" that life is "a population of RNA molecules (a quasi-species) which is able to self-replicate and evolve in the process." I will use the modified definition not only because I am one of those "adherents" but also because it provides a useful metric (RNA) for the size of primitive life. As Luisi noted, this definition implies the presence of an external source of energy and/or reactive nutrients to maintain the life. It also specifies the need for RNA but no other molecular species, but it is likely that some other organics were required.

Many scientists feel that this definition of life is inadequate because it does not require that this first life was protected from the vagaries of the environment on the primitive Earth by a surrounding compartment. This more complicated model of life was defined by Luisi as "a system which is spatially defined by a semipermeable component of its own making and which is self-sustaining by transforming external energy/nutrients by its own process of component production." Here I will also make the assumption that genetic information was also stored in RNA in this model of life. This more elaborate life-form may require additional biomolecules such as proteins for the synthesis of the monomers required for the biopolymers and the membrane.

It should be noted here that other biopolymers are also under consideration as either precursors to the RNA world or alternatives to it. For example, peptides have been synthesized in the laboratory (not by "prebiotic reactions") that are self-replicating by template-directed ligation (Lee et al., 1996; Severin et al., 1997).

A Summary of the Current State of Prebiotic Synthesis of RNA

The basic ingredients required for proposed models of primitive life are RNA, peptides, or proteins and membrane constituents. I recognize that the first life may not have utilized the types of biomolecules present in contemporary life, but at the present time there is very little information as to the possible structures of alternative life so the focus here will be on those molecules where information on possible prebiotic syntheses exists.

RNA

There has been progress in the understanding of the prebiotic synthesis of RNA monomers, but it is generally agreed there is much to be done to establish that these monomers were formed in sufficient amounts on the primitive Earth to lead to the formation of RNA oligomers. The research that has been done and those things that need to be accomplished were summarized (Ferris, 1987), and recent progress is reported by Zubay and coworkers (Zubay, 1994, 1998; Reimann and Zubay, 1999). Studies on a related ribopyranose-based RNA have been reported by Eschenmoser and coworkers (Pitsch et al., 1995).

It has been possible to catalyze the synthesis of RNA oligomers that contain up to 10 monomer units by the montmorillonite-catalyzed condensation of the 5'-phosphorimidazolides of 5'-nucleotides (ImpN; Figure 1; Ertem and Ferris, 1997; Kawamura and Ferris, 1994; Prabahar and Ferris, 1997). The RNAs are linked by 2', 5'- and 3',5'-phosphodiester bonds, pyrophosphate bonds and contain cyclic and linear oligonucleotides. Oligo(A)s as long as 50 mers have been prepared by the stepwise elongation of a decameric primer bound to montmorillonite by the daily addition of ImpA over a period of fourteen days (Figure 2; Ferris et al., 1996). This finding suggests that it may have been possible to form RNAs on mineral surfaces that were long enough to have served as templates for template-directed synthesis (Joyce and Orgel, 1993) and as catalysts for RNA ligation (Szostak and Ellington, 1993).

The replication of RNA, or any other genetic material, was a key process in the first life. It has not been possible to attain the non-enzymatic replication of RNA, but the template-directed synthesis of a complementary RNA has been demonstrated. Oligo(G)s over 40 mers in length are obtained in the template-directed reaction of 2-MeImpG on a poly(C) template (Figure 3; Inoue and Orgel, 1982). A less efficient template-directed synthesis is the formation of >6 mers of oligo(A)s by the reaction of

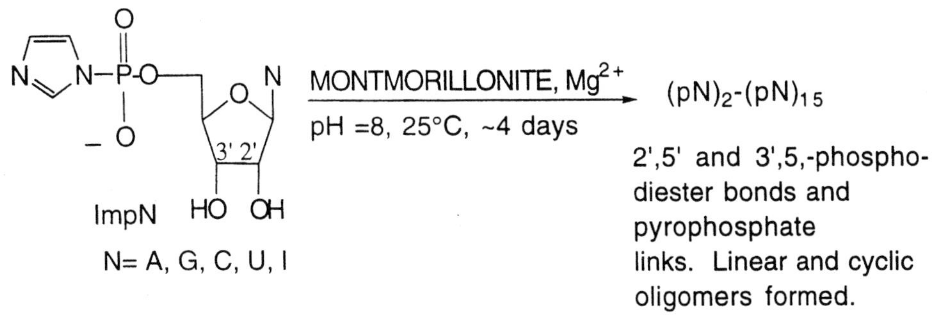

Figure 1. The montmorillonite-catalyzed oligomerization of activated mononucleotides.

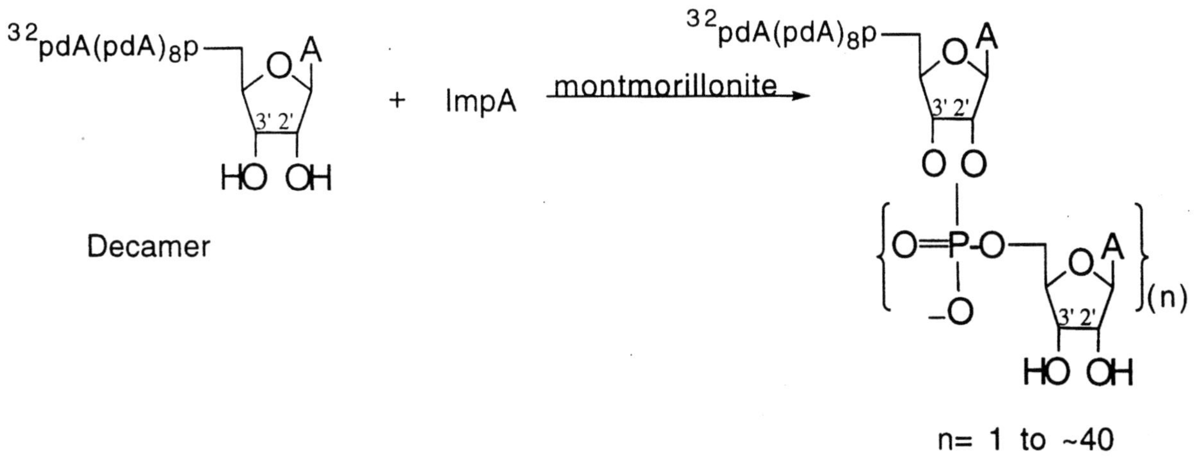

Figure 2. The montmorillonite-catalyzed elongation of a nucleic acid primer.

ImpA on a poly(U) template in the presence of Pb^{+2} (Sleeper et al., 1979). It has not been possible to demonstrate the non-enzymatic template-directed synthesis of pyrimidine oligomers on a polypurine nucleotide template or the template-directed synthesis on a heterogeneous template that contains more purine than pyrimidine nucleotides (Haertle and Orgel, 1986; Joyce and Orgel, 1986). It was observed that the heterogeneous RNAs formed in clay-catalyzed reactions, which contain 2'5'- and 3',5'-phosphodiester bonds, pyrophosphate links, and both cyclic and linear oligomers, do serve as templates for the synthesis of the complementary RNAs (Ertem and Ferris, 1997).

Prebiotic Syntheses of Polypeptides

There have been many reports of the prebiotic synthesis of short peptides in aqueous solution (for a brief summary see Liu and Orgel, 1998a), and it has been possible to make those that contain more than

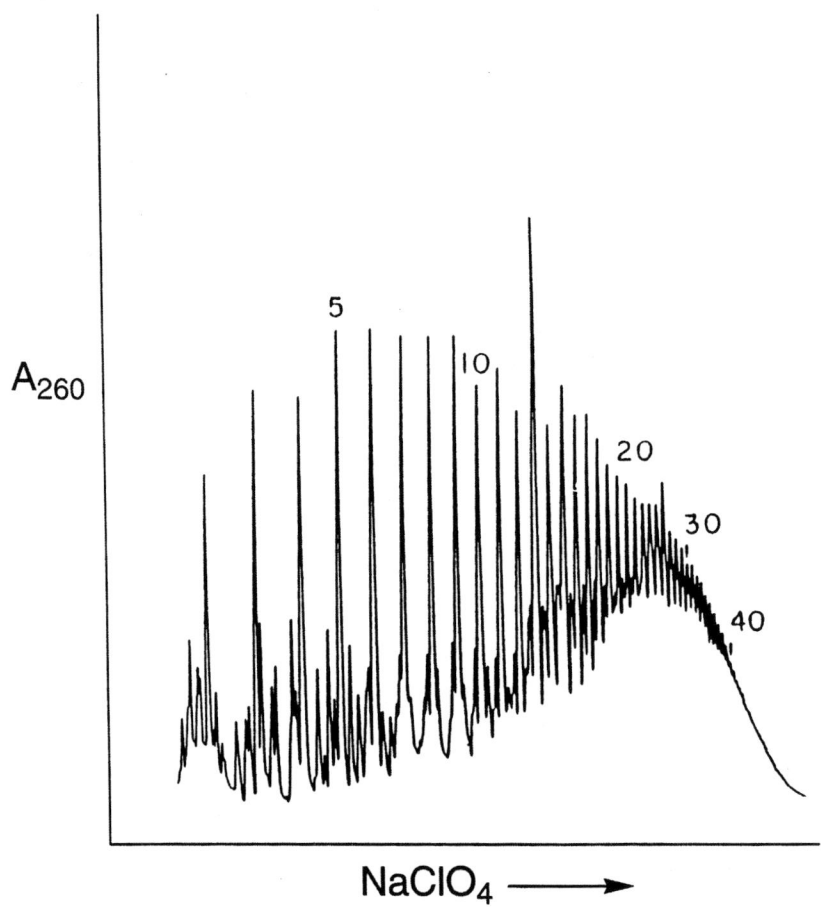

Figure 3. The template-directed synthesis of oligo(G)s from 2-MeImpG on poly(C) template. Reprinted, by permission, from T. Inoue and L.E. Orgel, *The Journal of Molecular Biology* (1982). Copyright © 1982 by Academic Press.

10 amino acids on mineral surfaces. Polymerization of the aminoacyladenylates of α-amino acids on montmorillonite yields polypeptides (Katchalsky, 1973; Paecht-Horowitz and Eirich, 1988; and previous papers in this series). Positively and negatively charged α-amino acids and β-amino acids form long chain polypeptides on mineral surfaces when carboxyl activating groups are added to the reaction mixture 20 to 50 times (Figure 4; Hill, Jr., et al., 1998; Liu and Orgel, 1998b). Hydroxyapatite, FeS_2, and the clay mineral illite were used in these studies. These findings suggest the possibility of the presence of catalytic polypeptides in the first life on Earth.

Prebiotic Membrane Formation

The formation of bilayer membranes requires the formation of a linear hydrocarbon chain containing greater than 10 to 12 carbon atoms with a charged or polar group on one end of the chain. No reports

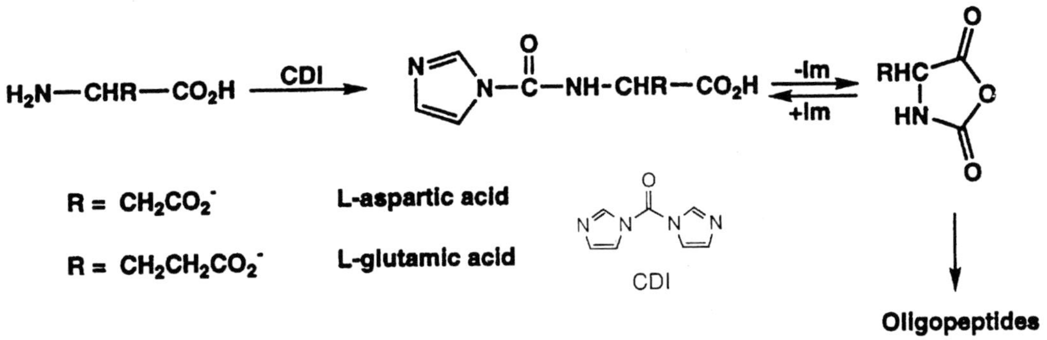

Figure 4. The elongation of a decameric glutamate and aspartate on hydroxyapatite and illite using carbonyl-diimidazole as the condensing agent. Reprinted, by permission, from *Origins of Life and Evolution of the Biosphere* by Aubrey Hill, Jr. et al. (1998). Copyright © 1998 by Kluwer Academic Publishers.

of the prebiotic synthesis of these long chains have been issued; however, their synthesis from formic and oxalic acids via the Fisher-Tropsch process in hydrothermal systems may be possible (Ferris, 1992; McCollum et al., 1999). The hydrolysis of esters and anhydrides of fatty acids results in their conversion to fatty acids, which associate into vesicles with diameters that range from 10 to 45 microns (Figure 5) (Walde et al., 1994). The presence of vesicles catalyzes the formation of additional vesicles as the hydrolysis proceeds.

An alternative source of vesicles may have been material brought to the primitive Earth by meteorites. A fraction isolated from the neutral extract of the Murchison meteorite forms vesicles capable of encapsulating a soluble dye during their formation in basic solution (Deamer and Pashley, 1989). The structure of this vesicle-forming material is not known, but it may be a carboxylic acid because these are one of the major constituents of the Murchison meteorite (Cronin et al., 1988)

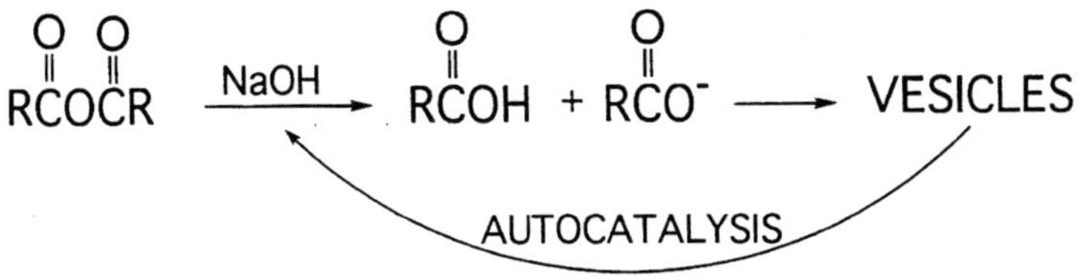

ANHYDRIDES OF CAPRYLIC (C_8) AND OLEIC (C_{18}) ACIDS

Figure 5. The formation of vesicles by hydrolysis of carboxylic acid anhydrides with base. The hydrolysis is catalyzed by the vesicles formed.

Size Estimates of the First Life

Life on the Rocks: A Minimal Form of Life

"Life on the rocks," a designation coined by Leslie Orgel, describes a system in which key processes for the formation of the biopolymers of life occur on mineral surfaces. The concept can be extended to a simple living system if the integrity of the life depends on the binding of molecules undergoing synthesis, replication, and mutation to the surface of a mineral assemblage. Such a system does not need to be bounded by a compartment to maintain its integrity.

In an RNA world on the rocks, the mineral-catalyzed synthesis of RNAs would generate RNAs capable of replication and evolution. The RNAs formed initially would have components capable of catalyzing these requisite functions. This would result in the preferential buildup of RNAs that carried out these essential tasks. Those RNAs that became detached from the mineral surface would initiate new centers of life when they became bound to other minerals. As the monomers for life are not synthesized on the mineral, this scenario requires the presence of a proximate source of activated monomers.

In the life on the rocks model there is no compartment surrounding the living system, so its size is determined by the size of the mineral assemblage that catalyzes the formation of the RNAs and the rates of synthesis and decomposition of the RNAs that are key to life processes.

Life Bounded by a Semipermeable Membrane

Life within a semipermeable membrane may be more resistant to changes in the environment, but it will also require the presence of a larger array of biomolecules than life on the rocks (Luisi, 1998). I assume that in the simplest case such life will require RNAs for the larger genome as well as RNAs to catalyze the synthesis of the RNA monomers and RNAs for synthesis and assembly of the semipermeable membrane (assumed to be constructed from fatty acids). Proteins may also have a role in this minimal life, but I will assume that randomly formed peptides and other biomolecules were adequate for the first simple life because this avoids the need for the translation machinery of protein synthesis.

The level of complexity of primitive life within a semipermeable membrane is comparable to the complexity of contemporary viruses. This is not to claim that the first life was a virus that evolved to a cell but rather that both are simple devices. The virus is able to replicate only with the aid of the biomolecules of a living cell. It has been proposed that viruses preserve a record of macromolecular evolution and may be molecular fossils of the RNA world (Maizels and Weiner, 1993). For an alternative view of both the RNA world and the thesis that RNA viruses contain vestigial RNA see Benner and Ellington (1987). Primitive life probably required the presence of preformed biomolecules that could be appropriated for its own purposes. Consequently, I have chosen virus as a metric for the measurement of the size of the first life. In Table 1 are listed size data on some RNA viruses with simple shapes so that their volume can be calculated assuming they are spheres. The Qβ virus, which has the 3 genes, is able to efficiently pack RNA within its membrane wall while the L-A virus has double-stranded RNA and presumably more protein and other biomolecules than does Qβ.

If it is assumed using Qβ as the model that a single-stand RNA gene has about 1,500 bases, then it is possible to estimate the volume required for a simple life-form with a variety of genes assuming the close packing present in the Qβ virus. One can use the same approach for calculating the radius of a primitive cell with double-stranded RNA and a larger proportion of other molecules by using the L-A virus as the model (Table 2). The radius of the particles based on the RNA content of the Qβ virus is less

Table 1 Dimensions and RNA Content of Some RNA Viruses[a]

	Base Pairs or Bases	Radius Inside Capsid (nm)[b]	Base Pairs or Bases per nm^3
Double-stranded			
Reoviruses	~22,000	25	0.34
L-A Virus	4,600	19	0.16
Single-stranded			
Polio Virus	6,100	10	1.5
Qβ Virus	4,600	8	2.1

[a]Data from Casjens (1997), Fraenkel-Conrat et al. (1988), and Metzler (1977).
[b]Not counting the phospholipid membrane, which is assumed to be 4 nm thick.

Table 2 Volume and Radii of Spherical Primitive Life Determined on the Basis of the RNA Packing in Qβ and L-A Viruses[a]

Genes	Qβ		L-A	
	Volume (nm^3)	Radius (nm)	Volume (nm^3)	Radius (nm)
1	715	5.5	9,580	13
3	2,150	8.0	28,700	19
5	3,580	9.5	48,900	23
7	5,000	11	67,000	25
10	7,150	12	95,800	28

[a]The calculation assumes a gene of 1,500 base pairs or bases as is the case for the Qβ virus. The radius used does not include that for the phospholipid membrane, which is assumed to be 4 nm. Proportional amounts of protein and other biomolecules present in the Qβ virus are also assumed to be present.

than one-half that of the L-A virus, and the available volume for RNA and other molecules is larger by the cube of the differences in the radii.

If it was possible to have membrane-bounded life based with a minimum of five genes (ligase, replicase, monomer synthase, fatty acid synthase, and membrane synthase ribozymes) and it had that RNA packing density of Qβ, then one would need a 3,580-nm^3 volume in addition to the volume of the surrounding membrane. As noted previously, proteins and other biomolecules were probably present as well. This volume would be 50,000 nm^3 if the contents of this simple life were more like the L-A virus.

How do these values correspond to the "nannobacteria" postulated for ALH84001? The tubular structures are said to be 20 to 100 nm in length (McKay et al., 1996). A crude approximation of the dimensions of these "nannobacteria" was made by measuring one of them (shown in Figure 6B of McKay et al., 1996). It is estimated to be 120 nm long and 10 nm in diameter. Correcting for a 4-nm phospholipid membrane layer, the internal dimensions are 112 nm long and 2 nm in diameter. Its internal volume is 350 nm^3, assuming it is a cylinder. Since the volume is a function of the square of the radius and the least reliable measure is the diameter of the "nannobacteria" in this photograph, the volume was calculated on the basis of a diameter of 14 nm and therefore an internal diameter of 6 nm. Here the volume is estimated to be 3,170 nm^3, a value close to that of 4 to 5 genes packed as they would be in the Qβ virus but only one-third of a gene if it were packed as it is in the L-A virus. The conclusion from this exercise is that primitive life may have been as small as large "nannobacteria" if there was an efficient mechanism for packing its RNA.

There are many concerns that can be raised with this approximation. First, a better estimate of the dimensions of these "nannobacteria" is needed. A small error in the diameter results in a major change in the calculated volume Consequently the compartment size for the first life may have been larger. Third, the genes for primitive life were probably shorter than 1,500 bases so would have required a smaller container. It is assumed that these factors more or less cancel out in these approximations.

The Signature of the First Life

It is unlikely that there will be a direct fossil record of the RNA world on the rocks because its structure is determined by the mineral assemblage to which it is attached. In addition, it is unlikely that this life would be recognized in ancient rock formations on Earth or Mars because it would have left behind few unique signatures. Circumstantial evidence for such life may be found if it has been established that the mineral is an efficient catalyst for the formation of an essential biopolymer. An exhaustive survey of potential mineral catalysts is required before undertaking such a search for these minerals in ancient rock formations. Such a search would be facilitated if there was data that suggested that primitive life provided conditions resulting in the deposition of a mineral that does not form under the usual environmental conditions. This was observed by the formation of apatite containing occluded organics in a banded iron formation in Greenland (Mojizsis et al., 1996). Unfortunately, even if such minerals were discovered they would not be a marker unique for establishing the former presence of life on the rocks since any type of living system may have initiated the formation of the marker crystals (Schopf, this volume, pp. 88-93).

The possibility of detecting primitive compartmentalized life on Mars is much higher than it is on Earth. If life did exist on Mars, the probability of its detection is higher there because it has no history of plate tectonics which would have destroyed most of these fossils by recycling the lithosphere. The principal challenge will be to distinguish these spherical microfossils from other small spherical objects of the same size.

Acknowledgments

Drs. Michael Gaffey, William Hagan, Jr., and Sandra Nierzwicki-Bauer provided helpful comment on an initial draft of this manuscript. The study was supported by NSF grant CHE-9619149, NASA grant NAG5-4557, and NASA NSCORT grant NAG5-7598.

References

Benner, S.A., and A.D. Ellington (1987). The last ribo-organism. *Nature* **329**: 295-296.

Casjens, S. (1997). Principles of virion structure, function and assembly. Pp 3-37 in *Structural Biology of Viruses*. W. Chiu, R.M. Burnett, and R.L. Garcea (eds.). New York: Oxford University Press.

Cronin, J.R., S. Pizzarello, and D.P. Cruickshank (1988). Organic matter in carbonaceous chondrites, planetary satellites, asteroids and comets. Pp. 819-857 in *Meteorites and the Early Solar System*. J.F. Kerridge and M.S. Matthews (eds.). Tuscon, Ariz.: University of Arizona Press.

Deamer, D.W., and R.M. Pashley (1989). Amphiphilic components of the Murchison carbonaceous chondrite: Surface properties and membrane formation. *Origins Life Evol. Biosphere* **19**: 21-38.

Ertem, G., and J.P. Ferris (1997). Template-directed synthesis using the heterogeneous templates produced by montmorillonite catalysis. A possible bridge between the prebiotic and RNA worlds. *J. Am. Chem. Soc.* **119**: 7197-7201.

Ferris, J.P. (1987). Prebiotic synthesis—problems and challenges. *Cold Spring Harbor Symposia on Quantitative Biology* **52**: 29-35.

Ferris, J.P. (1992). Marine hydrothermal systems and the origin of life: Chemical markers of prebiotic chemistry in hydrothermal systems. *Origins Life Evol. Biosphere* **22**: 109-134.

Ferris, J.P., A.R. Hill, Jr., R. Liu, and L.E. Orgel (1996). Synthesis of long prebiotic oligomers on mineral surfaces. *Nature* **381**: 59-61.

Fraenkel-Conrat, H., P.C. Kimball, and J.A. Levy (1988). *Virology 2*. Englewood Cliffs, N.J.: Prentice-Hall, p. 440.

Haertle, T., and L.E. Orgel (1986). The template properties of some oligodeoxynucleotides containing cytidine and guanosine. *J. Mol. Evol.* **23**: 108-112.

Hill, Jr., A.R., C. Böhler, and L.E. Orgel (1998). Polymerization on the rocks: Negatively-charged a-amino acids. *Origins Life Evol. Biosphere* **28**: 235-243, scheme 1 on p. 236.

Inoue, T., and L.E. Orgel (1982). Oligomerization of (guanosine 5'-phosphor)-2-methylimidazolide on poly(C): An RNA polymerase model. *J. Mol. Biol.* **162**: 201-217, fig. 3(c) on p. 207.

Joyce, G.F., and L.E. Orgel (1986). Non-enzymic template-directed synthesis on RNA random copolymers: poly(C,G) templates. *J. Mol. Biol.* **188**: 433-441.

Joyce, G.F., and L.E. Orgel (1993). Prospects for understanding the origin of the RNA World. Pp. 1-25 in *The RNA World*. R.F. Gesteland and J.F. Atkins (eds.). Cold Spring Harbor, New York: Cold Spring Harbor Laboratory Press.

Katchalsky, A. (1973). Prebiotic synthesis of biopolymers on inorganic templates. *Naturwiss.* **60**: 215-220.

Kawamura, K., and J.P. Ferris (1994). Kinetic and mechanistic analysis of dinucleotide and oligonucleotide formation from the 5'-phosphorimidazolide of adenosine on Na^+-montmorillonite. *J. Am. Chem. Soc.* **116**: 7564-7572.

Lee, D.H., J.R. Granja, J.A. Martinez, K. Severin, and M.R. Ghadiri (1996). A self-replicating peptide. *Nature* **382**: 525-528.

Liu, R., and L.E. Orgel (1998a). Polymerization of b-amino acids in aqueous solution. *Origins Life Evol. Biosphere* **28**: 47-60.

Liu, R., and L.E. Orgel (1998b). Polymerization on the rocks: b-amino acids and arginine. *Origins Life Evol. Biosphere* **28**: 245-257.

Luisi, P.L. (1998). About various definitions of life. *Origins Life Evol. Biosphere* **28**: 613-622.

Maizels, N., and A.M. Weiner (eds.) (1993). The genomic tag hypothesis: Modern viruses as molecular fossils of ancient strategies for genome replication. Pp. 577-602 in *The RNA World*. Cold Spring Harbor, New York: Cold Spring Harbor Laboratory Press.

McCollum, T.M., G. Ritter, and B.R.T. Simoneit (1999). Lipid synthesis under hydrothermal conditions. *Origins Life Evol. Biosphere* **29**: in press.

McKay, D.S., E.K. Gibson, Jr., K.L. Thomas-Kepra, C.S. Romanek, S.J. Clemett, X.D.F. Chillier, C.R. Maechling, and R.N. Zare (1996). Search for past life on Mars: Possible relic biogenic activity in martian meteorite ALH84001. *Science* **273**: 924-930.

Metzler, D.E. (1977). *Biochemistry*. New York: Academic Press, p. 1129

Mojizsis, S.J., G. Arrhenius, K.D. McKeegan, T.M. Harrison, A.P. Nutman, and C.R.L. Friend (1996). Evidence for life on Earth before 3,800 million years ago. *Nature* **384**: 55-59.

Paecht-Horowitz, M., and F.R. Eirich (1988). The polymerization of amino acid adenylates on sodium-montmorillonite with preadsorbed peptides. *Origins Life Evol. Biosphere* **18**: 359-387.

Pitsch, S., R. Krishnamurthy, M. Bolli, S. Wendeborn, A. Holzer, M. Minton, C. Lesuer, I. Schloenvogt, B. Jaun, and A. Eschenmoser (1995). Pyranose-RNA('p-RNA'): Base-pairing selectivity and potential to replicate. *Helv. Chim. Acta* **78**: 1621-1635.

Prabahar, K.J., and J.P. Ferris (1997). Adenine derivatives as phosphate-activating groups for the regioselective formation of 3',5'-linked oligoadenylates on montmorillonite: Possible phosphate-activating groups for the prebiotic synthesis of RNA. *J. Am. Chem. Soc.* **119**: 4330-4337.

Reimann, R., and G. Zubay (1999). Nucleoside phosphorylation: A feasible step in the prebiotic pathway to RNA. *Origins Life Evol. Biosphere* **29**: in press.

Severin, K., D.H. Lee, J.A. Martinez, and M.R. Ghadiri (1997). Peptide self-replication via template-directed ligation. *Chemistry-A European J.* **3**: 1017-1024.

Sleeper, H.L., R. Lohrmann, and L.E. Orgel (1979). Temple-directed synthesis of oligoadenylates catalyzed by Pb^{2+} ions. *J. Mol. Evol.* **13**: 203-214.

Szostak, J.W., and A.D. Ellington (1993). In vitro selection of functional RNA sequences. Pp. 511-533 in *The RNA World*. R.F. Gesteland and J.F. Atkins (eds.). Cold Spring Harbor, New York: Cold Spring Harbor Laboratory Press.

Walde, P., R. Wick, M. Fresto, A. Mangone, and P.L. Luisi (1994). Autopoietic self-replication of fatty acid vesicles. *J. Am. Chem. Soc.* **116**: 11649-11654.

Zubay, G. (1994). A feasible prebiotic pathway to the purines. *Chemtracts-Biochem. Mol. Biol.* **5**: 179-189.

Zubay, G. (1998). Studies on the lead-catalyzed synthesis of aldopentoses. *Origins Life Evol. Biosphere* **28**: 13-26.

CONSTRAINTS ON THE SIZES OF THE EARLIEST CELLS

Jack W. Szostak
Howard Hughes Medical Institute and Department of Molecular Biology
Massachusetts General Hospital

Abstract

Any discussion of constraints on the minimum size of simple, early cells must be based on speculative deductions about the structure of long-extinct ancestral life-forms. I first discuss reasons for thinking that early cells were surrounded by lipid membranes, and then explore the implications of such a structure. The physical properties of membranes are strongly influenced by their degree of curvature, which is related to vesicle size. Small vesicles of 50- to 100-nm diameter have properties which might, under suitable conditions, result in the establishment of a spontaneous cell cycle. Even such small cells could encapsulate a simple genome and cellular metabolism. I conclude that small early cells are a viable possibility. Given present uncertainties, it seems wise to be prepared to detect life-forms of a wide range of sizes.

Introduction: Early Cellular Life

It is important to distinguish between truly early cellular life and the last common ancestor of existing life. The structural and biochemical similarities of all existing branches of life point to a complex cellular structure for the last common ancestor, characterized by a DNA genome encoding at least several hundred and possibly several thousand genes, ribosome-catalyzed protein synthesis using the standard genetic code, membrane-surrounded cells with a wide range of protein transporters, and a complex metabolism supporting sugar, amino acid, nucleotide, cofactor, and lipid biosynthesis based on ATP synthesis from an electrochemical proton gradient (1). Such a cell would clearly have internal mechanisms for the control of basic processes such as cell growth and division. The last common ancestor is not, in its fundamentals, much simpler or even much different from current eubacterial or archaebacterial cells. We must look back much further in time to find simpler evolutionary precursors of such cells, and further still to find structures simple enough to have formed spontaneously by molecular self-assembly, yet complex enough to have evolved into life as we know it.

What might such ancestral forms have looked like? The arguments for early cells with an RNA genome and ribozymes as catalysts have been made many times (2) and will not be repeated here. However, the results of numerous recent experiments have confirmed the ability of ribozymes to catalyze a wide range of chemical transformations, including peptide and nucleotide synthesis (3-5). Given these results, it does not seem too unreasonable to postulate an intermediate between the earliest cells and the last common ancestor in which coded protein synthesis had not yet evolved, but which had evolved to a level of moderate complexity. Such a cell would be a membrane-bounded compartment containing a nucleic acid (RNA or DNA) genome that was transcribed to yield several hundred ribozymes that maintained a complex metabolism similar to the primary metabolism of modern cells (6). Perhaps the greatest uncertainty in regard to the plausibility of such a cell is a mechanism of membrane transport in the absence of complex coded proteins. Peptides or polyketides synthesized by sequential enzymatic steps (as cells still do today) may have played a key role in membrane-related processes.

Such a cell is still far too complex to be anything but the result of a long process of Darwinian evolution, starting from a much simpler ancestor. It is the ability of such an ancestral cell to evolve into more complex structures by Darwinian evolution that places the most severe constraints on its structure. I shall argue that, in addition to a genome that can be replicated with reasonable but not perfect accuracy, some form of compartmentation is required to enable Darwinian evolution. The ability to evolve is what distinguishes systems that are alive in a biologically relevant sense from prebiotic chemical systems and from other types of growth and propagation. Consideration of the simplest possible structures capable of evolution provides a framework for discussion of the question of the minimal size of such a structure.

Role of the Membrane Compartment

The function of the RNA (or RNA precursor) in our hypothetical progenitor cell is to provide a mechanism for the storage and replication of information in a form that is both heritable and mutable. The capacity for mutation allows the organism to explore new ways of adapting to its environment, while the heritability of such changes means that a selective advantage can be passed on to future generations. What then is the role of the membrane, or more generally of any form of compartmentation that places a boundary between the inside and the outside of a cell? For complex cells, a membrane-bound compartment is required for the co-localization of genes with gene products and metabolites. But for very simple cells, the idea of a membrane-bound compartment raises problems such as how nutrients can be imported, and how growth and cell division occur. Nevertheless, the membrane performs a subtle but critical function, which is to keep RNA molecules that are related by descent together, thus allowing natural selection to work. Because this function is so important, and so little appreciated, I will discuss it first to provide a rationale for the subsequent discussion of the properties of simple membrane-bound cells.

Perhaps the easiest way to understand this function is to consider what would happen in the absence of compartmentation. Imagine an initial population of RNA replicase molecules in free solution with activated monomers, but without any other RNA molecules present to complicate matters. Each replicase could copy any other RNA replicase that it happened to use as a template. If a mutant RNA replicase arose, with superior efficiency or accuracy, it would be better at replicating other RNA replicases, but would have no selective advantage for itself. Even worse, when it chanced to be replicated, its daughter molecules would diffuse away from each other, and thus could not even help each other preferentially.

In contrast, replicases that are replicating inside a growing and dividing membrane-bound compartment or vesicle would be capable of evolving. In this simple cellular system, each vesicle would contain some finite number of replicase molecules, which would use each other as templates for replication. Division would result in smaller vesicles, each containing a random subset of the replicase molecules from the parental cell. Through successive generations of such growth and division, the replicase molecules present in any one vesicle would tend, on average, to be more closely related by descent than replicase molecules in different vesicles. A mutant replicase that arose within such a system *would* have a selective advantage because it would be replicating its close relatives. During successive divisions, random segregation into daughter cells would eventually result in the formation of cells containing only the mutant replicase. Such cells would replicate their (mutant) genome more efficiently than cells containing only the parental replicase, and would eventually predominate in the population. Although selection for being a good template can occur in solution, selection for being a good replicase requires compartmentalization. Other new functions that favor propagation of the whole system could also evolve only in a compartmentalized system. The key to rapid and sustained evolution lies in the synergistic interaction between the molecules of inheritance and the molecules of compartmentation.

An independent reason for favoring a compartment-based cellular system is that this would allow for replicases to self-assemble from separate molecular segments. Many ribozymes can be assembled from a set of smaller RNA fragments; this attractive idea has the advantage that only relatively short segments need to be copied (7). However, to keep the segments together, some boundary, such as a membrane, would be needed.

Is a compartment absolutely required? In principle, selection for replicase activity could occur with dimers or higher multimers of a replicase, in which the various units would take turns acting as replicase and template; however, these ideas imply very long RNAs and require rather complex and unlikely dynamics such as partial strand separation so that newly copied material can re-fold into a replicase, while remaining attached to its template. For these reasons I do not favor the idea of a "living molecule," i.e., a replicating evolving molecule that exists in free solution. If the requirement for compartmentation is valid, the search for life should focus on cellular structures. On the other hand, evidence for life, or even critically important pre-biotic structures, might be found at either the molecular or cellular size scale.

If RNA without a compartment can't evolve (other than to be a better template), what about compartments without RNA? The beauty of membrane vesicles is that they are self-assembling structures that form spontaneously once a critical concentration of amphiphilic molecules exists. A variety of proposals have been made for the origin of life in self-sustaining metabolizing structures. Some such structures invoke networks of catalytic peptides or other molecules, while others postulate surface catalysis on colloidal particles of clay, FeS, or other materials. Autocatalytic networks are attractive from a theoretical perspective because they encode information in a distributed form, and can evolve by incorporating new catalytic processes. However, I suspect that the rarity of efficient catalysts in sequence space makes such models physically unrealistic. Vesicles containing catalytically active colloidal particles might develop quite complex chemistries, and may have been significant in the generation of monomers for the synthesis of RNA or its progenitors. Such particles could even be pre-biotic precursors of the first living cells. However, without a mechanism for heritable variation, they could not evolve in a Darwinian sense.

What about forms of compartmentation other than membranes? In principle, any medium that limits macromolecular diffusion more than small molecule diffusion, such as the interior of a gel matrix, or a micro or nano-porous rock, might suffice to keep molecules related by descent preferentially together. Fascinating experiments with Qβ replicase at least raise the possibility of such a mechanism (8). In such a scenario, it is hard to even say what the relevant size domain is. Another interesting possibility that has been suggested is an emulsion, in which small aqueous compartments in a non-aqueous matrix house replicating molecules (9). The variety of such possible systems emphasizes the importance of looking for life or its precursors in a wide range of environments. Nevertheless, since all present-day life consists of membrane-bound cells, any such precursor of life must at some point have made a transition to a membrane-bound compartment housing a replicating informational molecule. It is therefore worth considering if there are any significant size constraints on such a possible ancestral form of cellular life.

Size and the Cell Cycle

Membrane vesicles can be made from a wide range of phospholipids and other components, in a wide range of sizes. The vesicles that bud spontaneously from the surface of dried phospholipid films suspended in buffer tend to be large (1-10 μm) and multilamellar. However, when subjected to strong shear forces, either by sonication or by being forced through small pores under pressure, unilamellar vesicles as small as 50 nm in diameter are readily generated.

Arguments can be made in favor of either large or small vesicles as the most likely basis for early cellular life. The key challenge here is to come up with a plausible mechanism for cell growth and division, given that early primordial cells lacked all of the sophisticated internal machinery evolved by modern cells to control their growth and to mediate the physical process of division. Very small vesicles have an important property that may be relevant in this regard: because of their small size and strong curvature, the membrane is highly strained. The growth of such strained vesicles is thermodynamically favored by the relaxation that occurs as size increases and curvature decreases. Growth can occur spontaneously, either slowly by incorporation of additional lipid molecules, or rapidly by fusion with other vesicles. Incorporation of additional lipid can occur by transfer through solution from micelles or other small vesicles, and transfer is faster into smaller vesicles (10). Further studies of vesicle growth by this mechanism would be very useful in assessing this model for spontaneous growth. Vesicle fusion processes have been studied in much more detail (both as models of biological membrane fusion events and because of the potential importance of vesicles in drug delivery). Depending on the nature of the lipid, vesicle fusion can be mediated by Ca^{++}, by dehydration or by certain "fusogenic" peptides. Smaller vesicles tend to fuse much more readily than larger vesicles, although they are also less stable to phase changes in some conditions (11). Once larger vesicles have been generated, whether by growth or by fusion, they can divide into smaller vesicles with essentially no contents leakage, by shear-force-induced fission. In the laboratory, the simplest way of accomplishing this is by pressure-driven passage through small pores. The fact that vesicle growth and division can occur entirely under the influence of external environmental conditions raises the possibility of a primitive cell cycle, driven entirely by external physical forces, which is quite satisfying when considering a very simple cell that would lack all internal machinery for the control of growth and division. On the other hand, it is not clear what, if any, natural setting could provide the physical basis for the vesicle fission part of the cycle (passage of water in a hydrothermal vent system through microporous rock? wave action at the surface of a lake?). An interesting alternative possibility involves small vesicles formed spontaneously from short-chain lipids with a high intrinsic membrane curvature; synthesis of additional lipid by internal metabolic processes leads to vesicle growth and spontaneous (thermodynamically favored) fission (12).

Can we conceive of an analogous spontaneous cell cycle for larger vesicles? Large vesicles form spontaneously and fragment under mild shear forces. Here, however, it is the growth part of the cycle that is problematic, because of the absence of a thermodynamic driving force. Growth by incorporation of lipid molecules from solution seems unlikely (unless they are internally generated), and the fusion reactions of larger vesicles are less well studied. In general, concentrations of divalent cations that lead to the fusion of unstrained vesicles are very close to the concentrations that cause lipid phase changes and complete vesicle disruption (11). However, it must be emphasized that there have been very few studies of vesicle-based model systems for cell growth and division. One recent study suggests that hydrocarbons and single chain lipids may facilitate the fusion of larger vesicles (13). This observation raises the fundamental problem that, because the pre-biotic synthesis of amphiphilic molecules is so poorly studied, we have little idea of what kinds of molecules we should be looking at when studying model vesicle systems. Shorter lipid chains are known to generate less stable vesicles that are more permeable to small molecules such as nucleotides (14). Clearly, further studies of the physical properties of vesicles generated from pre-biotically plausible amphiphilic molecules would go a long way toward constraining the possible sizes of early cellular vesicles.

Very Small Cells Probably Transient

Although reasonable arguments can be made that the first cells might have been very small, it seems likely that such life-forms would have been quite transient, and soon superseded by the evolution of larger and more complex cells. Simple calculations based on volume suggest that a small vesicle could hold a maximum of about 1,000 medium-sized ribozymes (4- to 5-nm diameter, ~70 nucleotides), which, allowing for some redundancy, means that up to perhaps 100 distinct functions could be encoded. Thus, a very small (50-nm diameter) organism might in principle evolve to a level of moderate complexity without having to enlarge to the point of changing the basic physical phenomena involved in the cell cycle. However, complexity beyond this level would almost certainly require an increase in size. Since a vesicle of 50-nm internal diameter has a volume of ~60,000 nm^3 vs. 1 nm^3 for one base-pair in a duplex, the absolute upper limit on packing a double-stranded genome is about 60 kb, or 40 kb at a reasonable packing density. Even to hold a genome of 1 Mb (all free-living bacteria have genomes larger than this) would require a 150-nm diameter vesicle, and to keep the genome to less than 10% of total cell volume would require a minimum of a 300-nm diameter vesicle, or alternatively a 100-nm diameter cylinder of length 1,500 nm. Such a level of genomic complexity could easily accommodate considerable metabolic and structural complexity, including protein synthesis and internal regulation of cell growth, shape, and division. Although small relative to the size of most present-day microorganisms, this size may represent the lower limit necessary for organisms to maintain the complexity required to be competitive as a free-living life-form. Such life-forms would presumably have outcompeted and driven to extinction their smaller and simpler relatives, unless there were physiological factors or specialized ecological niches that favored the survival of small cells.

One obviously relevant physiological factor is that the high surface to volume ratio of small cells could help to compensate for the difficulties involved in transport of nutrients across membranes before the advent of protein transporters. Of course this is a two-edged sword, and the loss of essential metabolic intermediates would become a serious problem. This problem is exacerbated for small vesicles, since a single molecule in a 50-nm vesicle has a concentration of ~30 μM. A high-radiation environment might also initially favor small, simple cells, with a restricted genomic and cellular target size. However, the example of *M. radiodurans* shows that the evolution of repair functions can more than compensate for such environmental factors. As mentioned above, cell division of very small early cells would, at least initially, require an external source of energy in the form of an environment that provided very high shear forces. Although selection for the ability to grow outside such a restricted environment would be very strong, growth within that niche could be difficult or impossible for larger cells, at least until the evolution of rigid cell walls. Another niche that might be limited to very small cells could be nanoporous media such as, perhaps, compressed sediments. Such considerations may suggest that very small cells, if they ever existed, would be severely restricted in their distribution, both temporally and spatially.

Conclusions

Our present degree of knowledge is inadequate to strongly constrain the possible sizes of early cells. Vesicles as small as 50-nm diameter can be generated, could encapsulate small replicating informational polymers, and have at least some attractive properties in terms of the potential for a spontaneous cell cycle. Although such structures may not be the most likely form of primordial life, it would not be wise to ignore this possibility.

References

1. Benner, S.A., Ellington, A.D., and Tauer, A. (1989). *Proc. Natl. Acad. Sci. USA* **86,** 7054-7058.
2. Gilbert, W. (1986). *Nature* **319,** 618.
3. Zhang, B. and Cech, T.R. (1997). *Nature* **390,** 96-100.
4. Unrau, P.J. and Bartel, D.P. (1998). *Nature* **395,** 260-263.
5. Wilson, D. and Szostak, J.W. (1998). *Ann. Rev. Biochem.* **in press**.
6. Benner, S.A. and Ellington, A.D. (1987). *Nature* **329,** 295-296.
7. Doudna, J.A., Couture, S., and Szostak, J.W. (1991). *Science* **251,** 1605-1608.
8. Chetverina, H.V. and Chetverin, A.B. (1993). *Nucleic Acids Res.* **21,** 2349-2353.
9. Tawfik, D.S. and Griffiths, A.D. (1998). *Nat. Biotechnol.* **16,** 652-656.
10. Brown, R.E. and Hyland, K.J. (1992). *Biochemistry* **31,** 10602-10609.
11. Wilschut, J., Duzgunes, N., Hoekstra, D., and Papahadjopoulos, D. (1985). *Biochemistry* **24,** 8-14.
12. Schmidli, P.K., Schurtenberger, P., and Luisi, P.L. (1991). *J. Am. Chem. Soc.* **113,** 8127-8130.
13. Basanez, G., Goni, F.M., and Alonso, A. (1998). *Biochemistry* **37,** 3901-3908.
14. Chakrabarti, A.C., Breaker, R.R., Joyce, G.F., and Deamer, D.W. (1994). *J. Mol. Evol.* **39,** 555-559.

HOW SMALL CAN A MICROORGANISM BE?

Steven A. Benner
Departments of Chemistry and Anatomy and Cell Biology
University of Florida

Abstract

Much of the volume of a bacterial cell is filled with machinery (ribosomes) that converts information in the genetic biopolymer (DNA) into information in the catalytic biopolymer (protein). This places a limit on the size of a two-biopolymer living system that all but certainly excludes cells as small as (for example) the structures observed in the Allan Hills meteorite derived from Mars. Life that uses a single biopolymer to play *both* genetic and catalytic roles could conceivably fit within a smaller cell, however. No biopolymer has yet been found that can play both roles, and the chemical demands for genetics and catalysis are frequently contradictory. A catalytic biopolymer should have many building blocks; a genetic biopolymer should have few. A catalytic biopolymer should fold easily; a genetic biopolymer should not. A catalytic biopolymer must change its physical properties rapidly with few changes in its sequence; a genetic biopolymer must be COSMIC-LOPER (Capable Of Searching Mutation-space Independent of Concern over Loss of Properties Essential for Replication), with physical properties largely unchanged by changes in sequence. This article reviews the chemical plausibility of a single biopolymer that might make an effective compromise between these competing demands, and therefore permit life within very small cells.

Two-biopolymer Life-forms and One-biopolymer Life-forms

In terms of its macromolecular chemistry, life on Earth is a "two-biopolymer" system. Nucleic acid is the genetic biopolymer, storing information within an organism, passing it to its descendants, and suffering the mutation that makes evolution possible. Nucleic acids also direct the biosynthesis of the second biopolymer, proteins. Proteins generate most of the selectable traits, from structure to motion to catalysis. The two-biopolymer strategy evidently works well. It has lasted on Earth for billions of years, adapting to a remarkable range of environments, surviving formidable efforts by the cosmos to extinguish it, and generating intelligence capable of exploring beyond Earth.

The terrestrial version of two-biopolymer life contains a well-recognized paradox, however, one relating to its origins. It is difficult enough to envision a non-biological mechanism that would allow either proteins or nucleic acids to emerge spontaneously from non-living precursors. But it seems astronomically improbable that *both* biopolymers arose simultaneously *and* spontaneously, and even more improbable that both arose spontaneously, simultaneously, and as an encoder-encoded pair.

Accordingly, "single-biopolymer" models have been proposed for life that may have preceded the two-biopolymer system that we know on contemporary Earth (Joyce et al., 1987). Such models postulate that a single biopolymer can perform both the catalytic and genetic roles and undergo the Darwinian evolution that defines life (Joyce, 1994). RNA was proposed some time ago as an example of such a biopolymer (Rich, 1962; Woese, 1967; Orgel, 1968; Crick, 1968). This proposal became more credible after Cech, Altman, and their coworkers (Cech et al., 1981; Zaug and Cech, 1986; Guerrier-Takada et al., 1983) showed that RNA performs catalytic functions in contemporary organisms. The notion of an "RNA world," an episode in natural history when RNA served both genetic and catalytic roles, is now part of the culture of molecular biology (Watson et al., 1987).

Single-biopolymer Systems and Extraterrestrial Life

"Single-biopolymer" models for Darwinian chemistry have relevance to the search for extraterrestrial life. For example, some biologists have argued that the microstructures identified by McKay et al. (1996) in the Allan Hills meteorite, which are 20 to 100 nanometers across, are too small to be the remnants of living cells (Kerr, 1997). The argument is that the ribosome is 25 nm across, ribosomes are a requirement for life, and placing ribosomes (ca. four ribosomes across the short dimension of the "cell") in the cell would exclude virtually every other biomolecule.

This view is narrowly formulated. Ribosomes are a requirement for a two-biopolymer life-form, such as those known on contemporary Earth. If a single-biopolymer (such as RNA) can serve both genetic and catalytic functions, ribosomes are not required. A smaller cell may be sufficient to hold a single-biopolymer life-form.

How much smaller might the cells of a single-biopolymer life-form be (excluding parasitic cells)? Translation places demands upon the volume of a typical two-biopolymer cell. If we do not consider water, approximately half of the material inside an *E. coli* cell is ribosomes, tRNA, and mRNA (Lewin, 1985). Thus, a single-biopolymer cell can be half the size of a two-biopolymer cell simply by discarding the translation material. Of the remaining half of the dry weight of the intracellular contents of *E. coli*, aminoacyl tRNA synthetases, proteins that form transcription complexes, and proteins catalyzing amino acid biosynthesis are a major contributor. Together, biomolecules required to support translation comprise more than half of the soluble proteins that form the "core metabolism" encoded by the protogenome (Benner et al., 1993), the organism at the hypothetical threefold point joining Archaea, Eucarya, and Eukaryota in the universal tree of life.

Models can be built for a minimal metabolism that might be used by a single-biopolymer life-form. If that biopolymer is RNA, Figure 1 offers an autotrophic metabolism that involves fixation of carbon dioxide (4 catalysts, by analogy with the reductive tricarboxylic acid cycle), carbohydrate biosynthesis (5 catalysts exploiting cyanide-based couplings and aldol reactions), triphosphate generation (6 catalysts), nitrogen metabolism (3 catalysts), and nucleotide biosynthesis (28 catalysts, adopted directly from contemporary pathways). Ignoring the thermodynamics of this pathway (which are expected to be favorable under reducing conditions; see McCollom and Shock, 1997), this model sustains a single-biopolymer life-form with ca. 50 biocatalysts. Although additional macromolecules would undoubtedly be useful to biosynthesize membrane components, transport metal ions and cofactors, and participate in gene regulation, a plausible model for minimal single-biopolymer autotrophic life could almost certainly be limited to fewer than 100 macromolecules, less than 10% of the number found in a typical autotrophic two-biopolymer genome. If different types of catalysts can have the same size, a single-biopolymer life-form might fit within a cell having 5% the volume of a contemporary terrestrial bacterium. This implies that the microstructures in the martian meteorite might *not* be too small to be fossils of a single-biopolymer form of life. Conversely, if the meteorite structures are indeed fossils, then they almost certainly are fossils of an organism that used only a single biopolymer.

Does a Single Biopolymer Exist That Is Capable of Genetics *and* Catalysis?

This discussion suggests that the answer to the title question depends on the answer to the question: Does a single biopolymer exist that can robustly do both genetics and catalysis. In this discussion, we focus on RNA as the most highly regarded candidate for the single biopolymer.

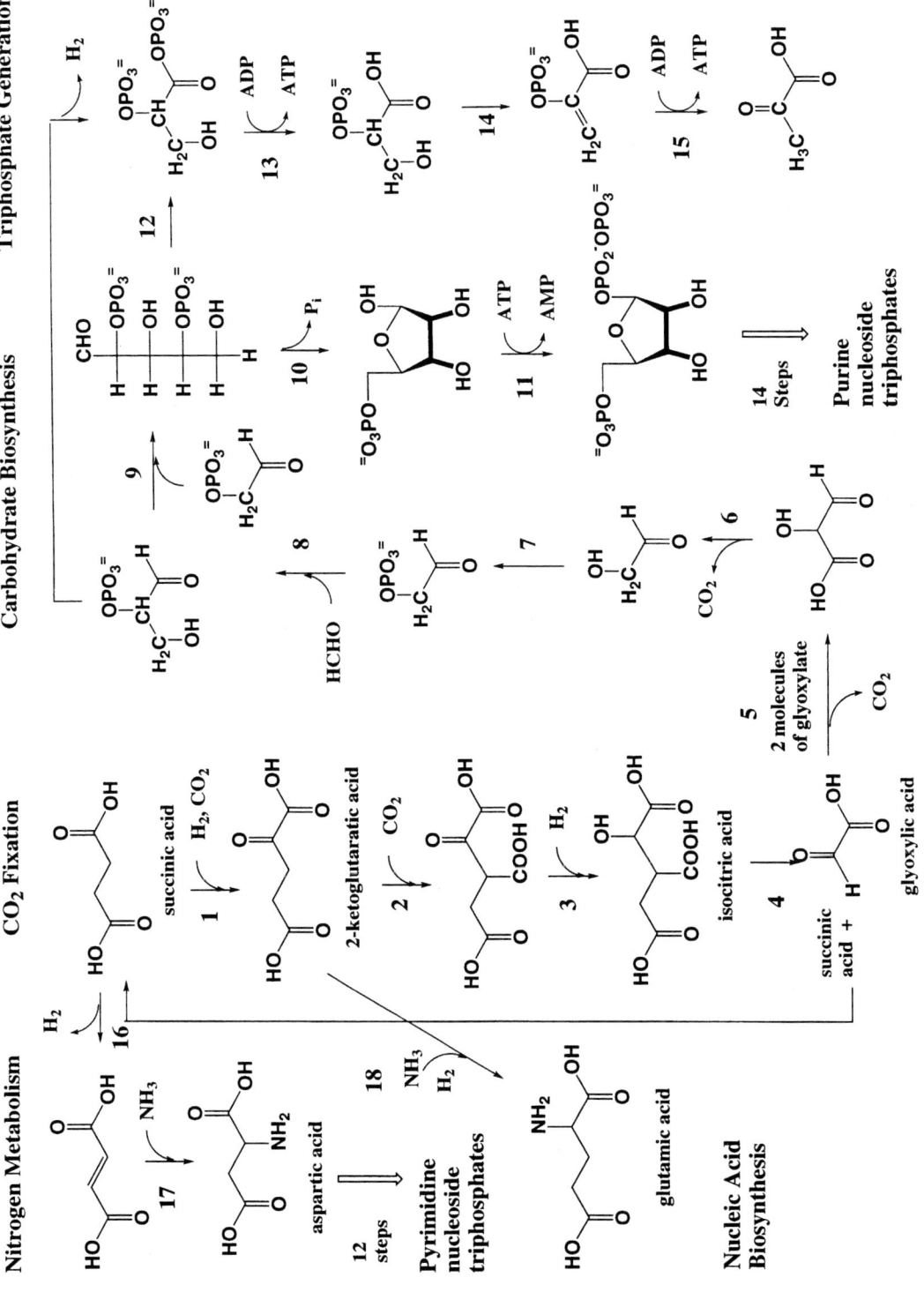

Figure 1. A hypothetical metabolism for a single-biopolymer autotrophic life-form based on RNA. Each reaction has a precedent in known chemistry, biological or non-biological. The driving force for the overall synthesis is not defined. The overall process would be exergonic under reducing conditions (see McCollum and Shock, 1997).

The Requirements for Genetics

A NASA workshop defined life as "a self-sustaining chemical system capable of undergoing Darwinian evolution" (Joyce, 1994). The genetic component of this definition is contained within the concept of Darwinian evolution. It includes not only the ability to be reproduced, but also the ability to survive mutation in a way that can create a change in phenotype that is selectable.

As discussed elsewhere (Benner and Switzer, 1998), many molecular systems can be reproduced and can form structures, catalysis, or other lifelike phenotypes. The most substantial challenge facing those attempting to develop a system that models life is to identify a biopolymer that can undergo mutation in a non-destructive way. Specifically, to support Darwinian evolution, a biopolymer must be able to search "mutation-space" independent of concern that it will lose properties essential for replication. If a substantial fraction of the mutations possible within a genetic information system cause a biopolymer to precipitate, unfold, or otherwise no longer be recognizable by the catalyst responsible for replication, then the biopolymer cannot evolve. We designate polymers that have this property as COSMIC-LOPER biopolymers (Capable of Searching Mutation-space Independent of Concern over Loss Of Properties Essential for Replication).

DNA and RNA are COSMIC-LOPER biopolymers. A mutant of a DNA sequence is as likely to dissolve in water, pair via Watson-Crick rules, template complementary strands, and be a substrate for DNA polymerases as its parent. The COSMIC-LOPER behavior is not absolute. If an RNA sequence wanders into a G-rich region of sequence space, it may become insoluble, or otherwise incapable of acting as a template. But these regions are exceptions.

Because of the familiarity of the "rule-based" molecular recognition properties displayed by DNA and RNA, the uniqueness of nucleic acids with respect to their COSMIC-LOPER behavior is often overlooked. In fact, very few classes of organic molecules can suffer changes in structures without significant changes in their physical properties. Perhaps the best example is proteins. The physical properties of proteins (including their solubility) can change dramatically upon point mutation within the mutation space allowed by the 20 standard amino acids. Again, there are many examples of this phenomenon in Nature (for example, hemoglobin in sickled cells). Designed peptides provide other examples. For example, altering their structure of a peptide designed to catalyze the decarboxylation of oxaloacetate by a single acetyl group changed substantially their level of aggregation, while altering their internal sequence at a single residue changes substantially their helicity (Johnsson et al., 1990, 1993). If solubility and/or helicity are essential to the replicatability of a peptide template, a large range of plausible mutation would destroy it. Protein is not COSMIC-LOPER, and is not expected to serve well as a genetic biopolymer, despite its acknowledged virtues as a catalytic biopolymer.

Starting in the 1980s, various groups altered the structure of nucleic acids to learn what structural features enable the rule-based molecular recognition properties (for a review, see Benner et al., 1998). The polyanionic nature of the oligonucleotide backbone appeared to be an important component of the COSMIC-LOPER behavior of nucleic acids; modifications of that backbone to remove the repeating charges created a biopolymer that no longer displayed rule-based molecular recognition (Richert et al., 1996).

Further, work expanding the number of letters in the genetic alphabet uncovered an intriguing relationship between the number of building blocks in a biopolymer and the fidelity of its synthesis. A genetic polymer should be replicated with a high (if not perfect) degree of fidelity. From both theory and experiment (Szathmary, 1992; Lutz et al., 1996), one expects higher fidelity with smaller genetic alphabets than large genetic alphabets.

The Requirements for Catalysis

Binding and catalysis (which may be viewed as binding to a transition state) require that the biopolymer present a series of specific interacting groups to the substrate. Here, diversity is advantageous. A case can be made that the 20 amino acid side chains found in natural proteinogenic amino acids provide a good sampling of the diversity that is available, in that it includes cationic groups, anionic groups, hydrophilic neutral groups, hydrophobic aliphatic groups, aromatic groups, and heterocycles, general acids and general bases, and nucleophilic groups. It has deficiencies. The standard 20 amino acids underrepresent heterocycles (compared, for example, with the U.S. Pharmacopoeia), it lacks a range of redox active side chains, and it is missing an electrophilic reactivity. But much of the diversity required for catalysis is present in standard proteins.

Catalysts must also surround a transition state, delivering contacting interactions from all sides. This, in turn, requires folding. Via a backbone with an equal number of hydrogen bond donors and acceptors, peptides fold well. Indeed, the feature most characteristic of proteins is that they precipitate (Benner, 1988b). Precipitation is folding, arising when the peptide prefers to interact with other peptides than with solvent. DNA and RNA in contrast, have a backbone of repeating negative charges. In the absence of cofactor (most commonly, divalent metal ion), there is no backbone-backbone interaction that supports the folding of an oligonucleotide (Richert et al., 1996).

The Contradicting Chemical Features Required for a Biopolymer That Does Both

This discussion makes evident that catalysis on one hand and genetics on the other place competing and contradictory demands on molecular structure. This implies in turn that it is difficult to find a single biopolymer that does both, suggesting that single-biopolymer life-forms might be less robust than two-biopolymer life-forms and that the small cells that single-biopolymer life enables might be scarcer in the universe than large cells. Let us review three specific contradictions:

1. A biopolymer specialized to be a catalyst must have many building blocks, so that it can display a rich versatility of chemical functionality required for catalysis. A biopolymer specialized for genetics must have few building blocks, as a way of ensuring faithful replication.
2. A biopolymer specialized to be a catalyst must fold easily so that it can form an active site. A biopolymer specialized for genetics should not fold easily, so that it can serve as a template (Richert et al., 1996).
3. A biopolymer specialized for catalysis must be able to change its physical properties rapidly with few changes in its sequence, enabling it to explore "function space" during divergent evolution. A biopolymer specialized for genetics must have physical properties largely unchanged even after substantial change in sequence (the COSMIC-LOPER property).

At the very least, a single-biopolymer attempting to support Darwinian evolution must reflect some sort of structural compromise between these goals. No fundamental principle guarantees that a polymeric system will make this compromise in a satisfactory way, however. The demands for functional diversity, folding, and rapid search of function space might be so stringent, and the demands for few building blocks, templating ability, and COSMIC-LOPER ability so stringent, that *no* biopolymer structure achieves a suitable compromise. Even if one exists, it may perform genetics and/or catalysis with poor robustness. Single-biopolymer life would then be fragile and easily extinguished. Life would be scarce in the universe because most of the initial forms would be driven to extinction before they could leap to a two-biopolymer structure. Conversely, if many-polymeric systems exist that make an acceptable compromise between the demands of catalysis and the demands of information storage, life would have emerged rapidly via single-biopolymer forms and be abundant in the universe in diverse forms.

Theoretical Evidence for a Robust Single-biopolymer System

Well before experiments were brought to bear on this problem, a theoretical argument was available that suggested that a single-biopolymer life-form might be possible. It began with three stipulations: (1) that life on Earth did not arise via divine intervention, (2) that spontaneous generation of a two-biopolymer system is not possible, but (3) that spontaneous generation of one biopolymer is possible. From the (obvious) fact that life exists on Earth, it can be concluded that a single biopolymer must have existed that performs both genetics and catalysis; this is the only way to explain the origin of life on Earth.

This proposal in one of various forms was made in the 1960s (Rich, 1962; Woese, 1967; Orgel, 1968; Crick, 1968). The extent to which the proposal begs questions was ameliorated by a rational analysis of contemporary biochemistry that began in the 1970s, when Usher and McHale (1976), White (1976), and Visser and Kellogg (1978) suggested that elements of contemporary metabolism (in particular, the structure of cofactors) might be viewed as vestiges of an "RNA world" (Gilbert, 1986). The emerging field of genomics was then used to generate internally consistent reconstructions for the ancient single-biopolymer life-forms. These reconstructions concluded from the abundance of its vestiges in modern metabolism that the RNA world was metabolically complex (Benner, 1988a; Benner et al., 1989; Benner et al., 1993). In modern metabolism, RNA fragments play roles for which they are *not* intrinsically suited. This suggests that these fragments originated during a time in natural history where RNA was the only available biopolymer, rather than by convergent evolution or recruitment in an environment where chemically better-suited biomolecules could be encoded. If the RNA world developed the RNA cofactors, ATP, coenzyme A, S-adenosylmethionine, and NADH, it follows that the RNA world needed these, presumably for phosphorylations, Claisen condensations, methyl transfers, and oxidation-reduction reactions (respectively).

These models imply that the RNA-based single-biopolymer life upon which all terrestrial life is founded had a complicated metabolism. This, in turn, implies that RNA *can* catalyze a wide variety of chemical reactions. This may be taken as indirect support for the existence of single-biopolymer life-forms, and from there, the possibility of very small cells.

The Experimental Evidence

These types of arguments, together with the discovery of RNA catalysis, made hopes high when Szostak (1988), Joyce (1989a,b), Gold (Irvine et al., 1991), and their coworkers introduced "in vitro selection" as a combinatorial tool to identify RNA molecules that catalyze specific reactions. If RNA was indeed as effective a catalyst as the reconstruction of the RNA world would imply, in vitro selection should rapidly generate the ultimate goal, an RNA (or DNA) molecule that catalyzes the template-directed polymerization of RNA (or DNA), a molecular system able to undergo Darwinian evolution. If selection procedures were appropriately designed, they should also produce RNA catalysts for almost any other reaction as well.

In contrast with these hopes (and only by this contrast), in vitro selection has been disappointing. RNA has proven to be an intrinsically poor matrix for obtaining catalysis, especially when compared with proteins. For example, to have a 50% chance of obtaining a single RNA molecule capable of catalyzing a template-directed ligation reaction by a modest (by protein standards) factor of 10,000, Bartel and Szostak (1993) estimated that one must sift through 2×10^{13} random RNA sequences 220 nucleotides in length. To obtain a catalyst with a factor of 10 greater catalytic power, one must increase the size of the library being searched by a factor of 1,000. This is poor catalysis, at least by comparison with proteins.

Although many laboratories have tried, only a few have managed to extend the scope of RNA catalysis beyond the phosphate transesterification reactions in which it was originally observed. For example, attempts to obtain an RNA catalyst for a Diels-Alder reaction using in vitro selection failed (Morris et al., 1994); the same reaction is readily catalyzed by protein antibodies (Gouverneur et al., 1993). Attempts to obtain RNA that catalyzes amide synthesis have succeeded, but with difficulty (Zhang and Cech, 1997; Wiegand et al., 1997). The fact that such successes came only after many attempts is indicative of a relatively poor catalytic potential in oligonucleotides.

The comparison with peptides is instructive. For example, short (14 amino acids) peptides accelerate the rate-determining step for the amine-catalyzed decarboxylation of oxaloacetate by more than three orders of magnitude (Johnsson et al., 1993), not far below the acceleration observed for the first-generation ligases observed in the Bartel-Szostak selection beginning with 10^{13} random RNA sequences. Further, the peptide is less than 10% the size of the RNA motif. Combinatorial experiments starting from this design (Perezpaya et al., 1996; Baltzer, 1998) suggested that perhaps only 10^7 random sequences must be searched to get a similar catalytic effectiveness as is observed in a library of 10^{13} RNA molecules. This suggests that peptides are intrinsically a millionfold fitter as catalysts than RNA.

The comparison is imperfect, of course, as it involves different reactions and different design strategies. This imperfection characterizes most of the comparisons that can be made at present. Not surprisingly, ribozymes are most frequently sought for reactions where oligonucleotides are most likely to be effective catalysts (for example, where oligonucleotides themselves are substrates), while peptide catalysts are most frequently sought for reactions suited for peptide catalysts (for example, those that make use of functional groups found on amino acid side chains). This makes the comparison non-quantitative, but useful nevertheless as an estimate of how well oligonucleotides and oligopeptides respectively perform when challenged by their favorite target reactions.

Biopolymers That Are Not (Exactly) RNA Or DNA

The failure of in vitro selection experiments with RNA to rapidly generate self-replicating systems challenges the notion that life emerged in a fashion directly analogous to the way in which in vitro selections are presently being done in the laboratory. This, in turn, means that these experiments failed to provide positive evidence that a single-biopolymer system exists, which in turn implies that we cannot confidently invoke a single-biopolymer life-form when we wish to argue that a very small structure (for example, on Mars) is a vestige of a primitive cell.

These experiments provided a direction, however. The apparent superiority of proteins as catalysts compared with RNA reflects (at the very least) the availability to proteins of a wider range of building blocks and catalytic functionality than in RNA. RNA lacks the imidazole, thiol, amino, carboxylate, and hydrophobic aromatic and aliphatic groups that feature so prominently in protein-based enzymes. RNA has only hydroxyl groups, polar aromatic groups, and phosphate groups. An uncounted number of studies with natural enzymes and their models has illustrated the use of this functionality by protein catalysts (Dugas, 1989).

Several groups are now seeking to add functionality to RNA and DNA. RNA might gain functionality using cofactors, much as contemporary proteins gain the functionality that they lack through vitamins. In a sense, this was already done in in vitro selection experiments, which nearly universally use the divalent magnesium cation, essentially as a cofactor. More recently, Breaker and his coworkers have expanded the approach to include organic molecules as second ligands in riboenzymes (Tang and Breaker, 1997).

A second solution was to append functionality to the standard nucleotides (Tarasow et al., 1997). Prompting this suggestion was the observation that contemporary tRNA and rRNA contain much of the functionality found in proteins but lacking in contemporary *encoded* RNA, including amino, carboxylate, and aliphatic hydrophobic groups (Limbach et al., 1994). These functional groups are introduced by post-transcriptional modification of encoded RNA. Some of these might even be placed by parsimony in the protogenome (Benner et al., 1989).

A third way to expand the functional diversity of nucleic acids is to increase the number of nucleotides in the nucleic acid alphabet. This can be done by using the non-standard hydrogen-bonding patterns permitted by the geometry of the Watson-Crick base pair (Switzer et al., 1989; Piccirilli et al., 1990). Additional letters in the genetic alphabet could carry a richer diversity of functionality. Indeed, one might imagine a new type of biopolymer, one carrying functionalization like proteins but able to be copied like nucleic acids (Kodra and Benner, 1997).

Conclusions

Each approach outlined above to increase the catalytic power of RNA as a single-biopolymer is only beginning to be explored. The title question will be answered only as this work proceeds. We believe that some of the most exciting results in chemistry in the next decade will come from efforts attempting to resolve the contradictions between catalysis and genetics in single-biopolymer systems in a way that will generate a biopolymer capable of both genetics and catalysis.

This question has implications for planetary exploration. The experiments with nucleic acid analogs has suggested as a hypothesis that a universal chemical characteristic of genetic biopolymers in water is a repeating charge, either an anion or a cation. This repeating charge may be both necessary and sufficient for COSMIC-LOPER behavior (Richert et al., 1996; Benner and Switzer, 1998). A repeating charge is a convenient biomarker for non-terrean genetic molecules. Future planetary probes might well search for such molecules.

Further, a single-biopolymer system should sustain work on Earth to learn how metabolic pathways might have emerged. In vitro selection permits would permit sequential selection for catalysts for individual metabolic steps (as shown in Figure 1). This would provide an experimental approach to identify the minimal cell, may generate new biomarkers, and could assist in the search for life on other planets.

Acknowledgments

We are indebted to NASA and the Office of Naval Research for supporting some of the work described here, and to the collaborators whose published work is cited.

References

Baltzer, L., 1998. Functionalization of designed folded peptides. *Opinions in Structural Biology* **8**, 466-470.
Bartel, D.P., and Szostak, J.W., 1993. Isolation of new ribozymes from a large pool of random sequences. *Science* **261**, 1411-1418.

Benner, S.A., 1988a. Reconstructing the evolution of proteins. Pps. 115-175 in *Redesigning the Molecules of Life*, Benner, S.A., ed. Heidelberg: Springer-Verlag.

Benner, S.A., 1988b. Evolution, physical organic chemistry, and understanding enzymes. *Adv. Clin. Enzymol.* **6**, 14-23.

Benner, S.A., 1989. Enzyme kinetics, and molecular evolution. *Chem. Rev.* **89**, 789-806.

Benner, S.A., and Switzer, C.Y., 1998. Chance, and necessity in biomolecular chemistry. Is life as we know it universal? In *Simplicity and Complexity in Proteins and Nucleic Acids*, Frauenhofer, H., Deisenhofer, J., and Wolynes, P.G., eds. Dahlem Workshop Report. Berlin: Dahlem University Press, in press.

Benner, S.A., Battersby, T.R., Kodra, J., Switzer, C.Y., Moroney, S.E., Vögel, J., MacPherson, L., von Krosigk, U., Hammer, C., Richert, C., Huang, Z., Horlacher, J., Schneider, K.C., König, M., Blättler, M., Arslan, T., Hyrup, B., Egli, M., Bäschlin, D., Müller, E., Schmidt, J., Piccirilli, J., Roughton, A., Held, H., Lutz, M., Eschgfäller, B., Jurczyk, S., Lutz, S., Hutter, D., Nambiar, K.P., and Stackhouse, J., 1998. Redesigning nucleic acids. *Pure Appl. Chem.* **70**, 263-266.

Benner, S.A., Cohen, M.A., Gonnet, G.H., Berkowitz, D.B., and Johnsson, K., 1993. Reading the palimpsest: Contemporary biochemical data and the RNA World. Pp. 27-70 in *The RNA World*, Gesteland, R., and Atkins, J., eds. New York: Cold Spring Harbor Laboratory Press.

Benner, S.A., Ellington, A.D., and Tauer, A., 1989. Modern metabolism as a palimpsest of the RNA World. *Proc. Natl. Acad. Sci. USA* **86**, 7054-7058.

Cech, T.R., Zaug, A.J., and Grabowski, P.J., 1981. In vitro splicing of the ribosomal RNA precursor of Tetrahymena. Involvement of a guanosine nucleotide in the excision of the intervening sequence. *Cell* **27**, 487-496.

Crick, F.H.C., 1968. The origin of the genetic code. *J. Mol. Biol.* **38**, 367-379.

Dugas, H., 1989. *Bioorganic Chemistry*, 2nd Ed. New York: Springer-Verlag.

Gilbert, W., 1986. The RNA world. *Nature* **319**, 618.

Gouverneur, V.E., Houk, K.N., Depascualteresa, B., Beno, B., Janda, K.D., and Lerner, R.A., 1993. Control of the exopathway, and endo-pathway of the Diels-Alder reaction by antibody catalysis. *Science* **262**, 204-208.

Guerrier-Takada, C., Gardiner, K., Marsh, T., Pace, N., and Altman, S., 1983. The RNA moiety of RNase P is the catalytic subunit of the enzyme. *Cell* **35**, 849-857.

Irvine, D., Tuerk, C., and Gold, L., 1991. Selexion. Systematic evolution of ligands by exponential enrichment with integrated optimization by non-linear analysis. *J. Mol. Biol.* **222**, 739-761.

Johnsson, K., Allemann, R.K., and Benner, S.A., 1990. Designed enzymes. New peptides that fold in aqueous solution, and catalyze reactions. Pp. 166-187 in *Molecular Mechanisms in Bioorganic Processes*, Bleasdale, C., and Golding, B.T., eds. Cambridge: Royal Society of Chemistry.

Johnsson, K., Allemann, R.K., Widmer, H., and Benner, S.A., 1993. Synthesis, structure, and activity of artificial, rationally designed catalytic polypeptides. *Nature* **365**, 530-532.

Joyce, G.F., 1989a. Amplification, mutation, and selection of catalytic RNA. *Gene* **82**, 83-87.

Joyce, G.F., 1989b. Building the RNA world. Evolution of catalytic RNA in the laboratory. *UCLA Symp. Mol. Cell. Biol. New Ser.* **94**, 361-371.

Joyce, G.F., 1994. Forward in *Origins of Life: The Central Concepts*, Deamer, D.W., and Fleischaker, G.R., eds. Boston: Jones and Bartlett.

Joyce, G.F., Schwartz, A.W., Miller, S.L., and Orgel, L.E., 1987. The case for an ancestral genetic system involving simple analogs of the nucleotides. *Proc. Natl. Acad. Sci. USA* **84**, 4398-4402.

Kerr, R.A., 1997. Ancient Life on Mars? Putative Martian microbes called microscopy artifacts. *Science* **278**, 1706-1707.

Kodra, J., and Benner, S.A., 1997. Synthesis of an N-alkyl derivative of 2'-deoxyisoguanosine. *Syn. Lett.*, 939-940.

Lewin, B., 1985. *Genes II*. New York: John Wiley & Sons, p. 95.

Limbach, P.A., Crain, P.F., and McCloskey, J.A., 1994. Summary. The modified nucleosides of RNA. *Nucl. Acids Res.* **22**, 2183-2196.

Lutz, M.J., Held, H.A., Hottiger, M., Hübscher, U., and Benner, S.A., 1996. Differential discrimination of DNA polymerases for variants of the non-standard nucleobase pair between xanthosine, and 2,4-diaminopyrimidine, two components of an expanded genetic alphabet. *Nucl. Acids Res.* **24**, 1308-1313.

McCollum, T.M., and Shock, E.L. 1997. *Geochim. Cosmochim. Acta.* **61**, 4375-4391.

McKay, D.S., et al., 1996. *Science* **273**, 924-930.

Morris, K.N., Tarasow, T.M., Julin, C.M., Simons, S.L., Hilvert, D., and Gold, L., 1994. Enrichment for RNA molecules that bind a Diels-Alder transition-state analog. *Proc. Natl. Acad. Sci. USA* **91**, 13028-13032.

Orgel, L.E., 1968. Evolution of the genetic apparatus. *J. Mol. Biol.* **38**, 381-393.

Perezpaya, E., Houghten, R.A., and Blondelle, S.E., 1996. Functionalized protein-like structures from conformationally defined synthetic combinatorial libraries. *J. Biol. Chem.* **271**, 4120-4126.

Piccirilli, J.A., Krauch, T., Moroney, S.E., and Benner, S.A., 1990. Extending the genetic alphabet: Enzymatic incorporation of a new base pair into DNA, and RNA. *Nature* **343**, 33-37.

Rich, A., 1962. On the problems of evolution, and biochemical information transfer. Pp. 103-126 in *Horizons in Biochemistry*, Kasha, M., and Pullman, B., eds. New York: Academic Press.

Richert, C., Roughton, A.L., and Benner, S.A., 1996. Nonionic analogs of RNA with dimethylene sulfone bridges. *J. Am. Chem. Soc.* **118**, 4518-4531.

Switzer, C.Y., Moroney, S.E., and Benner, S.A., 1989. Enzymatic incorporation of a new base pair into DNA, and RNA. *J. Am. Chem. Soc.* **111**, 8322-8323.

Szathmary, E., 1992. What is the optimum size for the genetic alphabet? *Proc. Natl. Acad. Sci. USA* **89**, 2614-2618.

Szostak, J.W., 1988. Structure, and activity of ribozymes. Pp. 87-114 in *Redesigning the Molecules of Life*, Benner, S.A., ed. Heidelberg: Springer-Verlag.

Tang, J., and Breaker, R.R., 1997. Rational design of allosteric ribozymes. *Chem. Bio.* **4**, 453-459.

Tarasow, T.M., Tarasow, S.L., and Eaton, B.E., 1997. RNA-catalyzed carbon-carbon bond formation. *Nature* **389**, 54-57.

Usher, D.A., and McHale, A.H., 1976. Hydrolytic stability of helical RNA: A selective advantage for the natural 3',5'-bond. *Proc. Natl. Acad. Sci. USA* **73**, 1149-1153.

Visser, C.M., and Kellogg, R.M., 1978. Biotin. Its place in evolution. *J. Mol. Evol.* **11**, 171-178.

Watson, J.D., Hopkins, N.H., Roberts, J.W., Steitz, J.A., and Weiner, A.M., 1987. *Molecular Biology of the Gene*, 4th Ed., p. 1115. Menlo Park, Calif.: Benjamin Cummings.

White III, H.B., 1976. Coenzymes as fossils of an earlier metabolic state. *J. Mol. Evol.* **7**, 101-104.

Wiegand, T.W., Janssen, R.C., and Eaton, B.E., 1997. Selection of RNA amide synthases. *Chem. Biol.* **4**, 675-683.

Woese, C.R., 1967. *The Genetic Code. The Molecular Basis for Genetic Expression.* New York: Harper & Row.

Zaug, A.J., and Cech, T.R., 1986. The intervening sequence RNA of Tetrahymena is an enzyme. *Science* **231**, 470-475.

Zhang B.L., and Cech T.R., 1997. Peptide bond formation by in vitro selected ribozymes. *Nature* **390**, 96-100.

Appendixes

Appendix A

Steering Group Biographies

Co-chair, Andrew H. Knoll—Dr. Knoll is a professor of biology at the Botanical Museum at Harvard University. His areas of expertise include the evolution of life, the evolution of Earth surface environments, and the relationships between the two. He is particularly interested in Archean and Proterozoic paleontology, carbonate sedimentology, and biogeochemistry. His current areas of research include both the early evolution of life and the neo-Proterozoic-Cambrian diversification of animals. Dr. Knoll is currently a member of the Space Studies Board and the 1998 NAS Nominating Committee. His previous NAS service includes membership on the Board on Earth Sciences and Resources, Panel on the Effects of Past Global Change on Life, Panel on the History of Life, Board on Earth Studies, Life Sciences Task Group, and the Committee on Planetary Biology and Chemical Evolution. Dr. Knoll has also served as a member of the delegation to the 30th International Geological Congress in Beijing and the 1997 Charles Doolittle Walcott Medal Selection Committee. He is a member of the American Philosophical Society and the American Academy of Arts and Sciences. Dr. Knoll received a B.A. from Lehigh University and an M.A. and a Ph.D. in geology from Harvard University.

Co-chair, Mary Jane Osborn—Dr. Osborn is a professor and is Head of Microbiology at the University of Connecticut Health Center. Her current research interests include the biogenesis of bacterial membranes. Dr. Osborn has served on numerous distinguished committees, including the National Science Board (80-86), the President's Committee on the National Medal of Sciences (81-82), the Advisory Council of the National Institutes of Health's Division of Research Grants (89-94; chair, 92-94), the Advisory Council of the Max Planck Institute of Immunobiology (74-78), the Board of Scientific Advisors for the Roche Institute for Molecular Biology (81-85; chair, 83-85), and the Governing Board of the National Research Council (90-93). She is a member of the National Academy of Sciences, the American Association for the Advancement of Science, the American Society of Biochemistry and Molecular Biology (president, 81-82), the American Chemical Society (chair, Division of Biological Chemistry, 75-76), the American Academy of Arts and Sciences (fellow; Council, 88-92), the Federation of American Societies for Experimental Biology (president, 82-83), the American Society for

Microbiology, and the American Academy of Microbiology. Dr. Osborn received a B.A. from the University of California at Berkeley, and a Ph.D. (biochemistry) from the University of Washington.

Norman R. Pace—Dr. Pace is a professor in the Department of Plant and Microbial Ecology, University of California at Berkeley. He is an internationally recognized expert in nucleic acids and enzymes. His studies of ribosomal RNA structures have set new standards for the definition of phylogenetic relationships among organisms. Dr. Pace has held academic positions with several universities, including the University of Colorado Medical Center, the National Jewish Hospital and Research Center, the University of Colorado Medical Center, and Indiana University. His research interests include RNA enzymes, RNA processing, macromolecular structure, molecular evolution, and microbial ecology. He has served on the editorial board of the *Journal of Biological Chemistry* and the journal *RNA* (of the RNA Society). He served as a member of the NRC's Committee on Planetary Biology, Committee on Chemical Evolution, Committee on Planetary and Lunar Exploration, and the Mars Rover/Sample Return Advisory Committee. Currently, Dr. Pace is a member of the Board on Scientific Counselors for the National Center for Biotechnology Information, National Library of Medicine. He is a fellow of the American Association for the Advancement of Science and of the American Academy of Microbiology. He is a member of the American Society for Microbiology. Dr. Pace received a B.A. (honors) from Indiana University and a Ph.D. from the University of Illinois at Urbana-Champaign.

John Baross—Dr. Baross, a professor of oceanography at the University of Washington, specializes in the ecology, physiology, and taxonomy of microorganisms from hydrothermal vent environments, as well as the use of biochemical and molecular methods to detect, quantify, and classify the same. Dr. Baross has particular interests in the microbiology of extreme environments and in the significance of submarine hydrothermal vent environments for the origin and evolution of life. He is a member of the American Society for Microbiology, the American Chemical Society, the Oceanography Society, the American Geophysical Union, the American Society for Limnology and Oceanography, and the Society for Industrial Microbiology (Puget Sound Branch). Dr. Baross is a former member of the National Research Council's Ad Hoc Task Group on Planetary Protection and of the American Academy of Microbiology Committee on the Future of Microbiology; a member of the Ridge Inter-Disciplinary Global Experiments (RIDGE) Steering Committee and of the RIDGE Observatory Coordinating Committee; and an advisory member for Europa Ocean Studies. Dr. Baross received a B.S. degree from San Francisco State University and a Ph.D. degree in marine microbiology from the University of Washington.

Mitchell Sogin—Dr. Sogin is the director of the Bay Paul Center for Comparative Molecular Biology and Evolution at the Marine Biological Laboratory at Woods Hole, Massachusetts. Dr. Sogin's research emphasizes molecular phylogeny and the evolution of eukaryotic ribosomal RNAs. He is a member of the American Society for Microbiology, the Society of Protozoologists, the International Society of Evolutionary Protozoologists, the Society for Molecular Biology and Evolution, the American Association for the Advancement of Science, and the American Society for Cell Biology. Dr. Sogin is an associate fellow of the Canadian Institute for Advanced Research, a division lecturer for the American Society for Microbiology, a recipient of the Stoll Stunkard Award from the American Society of Parasitologists, a fellow of the American Academy of Microbiology, a fellow of the American Association of Arts and Sciences, and a visiting Miller Research Professor at the University of California at Berkeley. He also serves on several editorial boards in his specialization. Dr. Sogin received B.S., M.S., and Ph.D. degrees in microbiology and molecular biology from the University of Illinois at Urbana.

Howard C. Berg —Dr. Berg is a professor in the Department of Molecular and Cellular Biology and Department of Physics at Harvard University and is a member of the Rowland Institute for Science. His primary research interest is the motile behavior of bacteria, including chemotaxis and flagellar rotation in *Escherichia coli*, flagellar rotation in a motile Streptococcus spp., and swimming of a novel cyanobacterium. His earlier work includes studies of spin exchange in the hydrogen maser, development of methods for sedimentation field flow fractionation, and studies of the architecture of the human erythrocyte membrane. His honors include a Fulbright Fellowship; membership in the Harvard Society of Fellows, an NSF Science Faculty Professional Development Award; and the Biological Physics Prize of the American Physical Society (with E.M. Purcell). He is a fellow of the American Academy of Microbiology and of the American Physical Society. He served on the Ford Foundation's Minority Review Panel on Biological Sciences and the NRC's Board on Physics and Astronomy. Dr. Berg received a B.S. degree in chemistry from the California Institute of Technology and a Ph.D. degree in chemical physics from Harvard University.

Appendix B

Request from NASA

National Aeronautics and
Space Administration
Headquarters
Washington, DC 20546-0001

MAR 13 1998

Reply to Attn of: SR

Dr. Claude Canizares
Space Studies Board
National Academy of Sciences
2101 Constitution Avenue, N.W.
Washington, DC 20418

Dear Dr. Canizares:

The investigation of the ancient Martian meteorite ALH84001 for evidence of life has revealed and highlighted a large degree of uncertainty in the possible existence of extremely small microorganisms here on Earth. The question of minimal microbial size is debated within the scientific community and there is no widely accepted theoretical minimum size for microorganisms. The Space Studies Board (SSB) has been a primary group advising NASA and seeks the Space Studies Board's advice concerning the size limits to life.

NASA has an interest in determining the current state of knowledge of the size limits to life on Earth, for microorganisms past and present, and in the theoretical limits. To understand the origin, evolution and distribution of life in the universe, a fundamental understanding of the size limits to life is needed. Scientists need to know what to look for and how to interpret the results. Considering the infancy of this research area and the continual advance in laboratory techniques, NASA also seeks recommendations concerning fertile research directions to explore the size limits to life.

In addition to considering the size below which life is not possible, there are related issues that should be addressed. What is the relationship between minimal size and the environment? Are very small microorganisms primitive or is their size a derived characteristic? What are the implications to the search for life on other planets?

Your help in addressing the questions about the size limits to life is greatly appreciated. Dr. Michael Meyer will be working with you and the SSB staff to finalize a Statement of Task for this study effort. Please contact him (202-358-0307) if you need further information about this request.

Sincerely,

Wesley T. Huntress, Jr.
Associate Administrator for
  Space Science

cc:
S/Dr. C. Pilcher
S/Dr. G. Soffen
S/Dr. E. Weiler
SR/Dr. M. Meyer
NRC/Dr. Alexander

Appendix C

Workshop Agenda

THURSDAY, OCTOBER 22, 1998

General Session

8:30 a.m.	Welcome and Opening Remarks	Andrew Knoll, Co-chair *Harvard University*
8:40 a.m.	E. William Colglazier, *Executive Officer* *National Research Council*	
9:00 a.m.	Edward Weiler, *Associate Administrator (acting)* *Office of Space Science, NASA*	
9:20 a.m.	Rita R. Colwell, *Director* *National Science Foundation*	
9:40 a.m.	Overview of the Workshop	Mary Jane Osborn, Co-chair *University of Connecticut Health Center*

Panel Sessions

9:50 a.m.	**PANEL 1**	Christian de Duve, Moderator *Christian de Duve Institute of Cellular Pathology*

What features of biology characterize microorganisms at or near nanometer scale?
Is there a theoretical size limit below which free-living organisms cannot be viable?
If we relax the requirement that cells have the biochemical complexity of modern cells, can we model primordial cells well enough to estimate their likely sizes?

10:00 a.m. Dan Fraenkel, *Harvard Medical School*
10:20 a.m. Jeffrey Lawrence, *University of Pittsburgh*

10:40 a.m. **Break**   Lecture Room

10:55 a.m. Monica Riley, *Woods Hole Marine Biological Laboratory*
11:10 a.m. David Boal, *Simon Fraser University*
11:30 a.m. Peter Moore, *Yale University*

12:00 noon **Lunch**

1:00 p.m.   **Panel 1 Discussion**   Christian de Duve, Moderator
                                                                   Christian de Duve Institute of Cellular Pathology
2:30 p.m.   **Concluding Remarks for Panel 1**   Christian de Duve, Moderator

2:40 p.m.   **Break**   Lecture Room

2:50 p.m.   **PANEL 2**   Ken Nealson, Moderator
                                                                   Jet Propulsion Laboratory

Is there a relationship between minimum size and environment?
Is there a continuum of size and complexity that links conventional bacteria to viruses?
What is the phylogenetic distribution of very small bacteria?

3:00 p.m.   James Van Etten, *University of Nebraska at Lincoln*
3:20 p.m.   Olavi Kajander, *University of Kuopio*
3:40 p.m.   Don Button, *University of Alaska at Fairbanks*
4:10 p.m.   James Staley, *University of Washington*
4:30 p.m.   Karl Stetter, *Universität Regensburg*

FRIDAY, OCTOBER 23, 1998

General Session

8:30 a.m.   Opening Remarks   Andrew Knoll, Moderator
                                                                   Harvard University

Panel Sessions

8:40 a.m.   **PANEL 2** (continued)   Ken Nealson, Moderator
                                                                   Jet Propulsion Laboratory

8:50 a.m.	Michael Adams, *University of Georgia*	
9:10 a.m.	Edward DeLong, *Monterey Bay Aquarium Research Institute*	
9:30 a.m.	**Panel 2 Discussion**	Ken Nealson, Moderator *Jet Propulsion Laboratory*
10:30 a.m.	**Break**	Lecture Room
10:45 a.m.	**Panel 2 Discussion** (Continued)	Ken Nealson, Moderator *Jet Propulsion Laboratory*
11:50 a.m.	**Concluding Remarks for Panel 2**	Ken Nealson, Moderator
12:00 noon	**Lunch**	
1:00 p.m.	**PANEL 3**	Andrew Knoll, Moderator *Harvard University*

Can we understand the processes of fossilization and inorganic chemistry sufficiently well to differentiate fossils from artifacts in a sample?

1:10 p.m.	William Schopf, *University of California at Los Angeles*	
1:30 p.m.	Jack Farmer, *Arizona State University*	
1:50 p.m.	John Bradley, *MVA, Inc.*	
2:10 p.m.	**Panel 3 Discussion**	Andrew Knoll, Moderator *Harvard University*
3:10 p.m.	**Concluding Remarks for Panel 3**	Andrew Knoll, Moderator
3:20 p.m.	**Break**	
3:30 p.m.	**PANEL 4**	Leslie Orgel, Moderator *Salk Institute*

Does our current understanding of the processes that led from chemical to biological evolution place constraints on the size of early organisms? If size is not constrained, are there chemical signatures that might record the transition to living systems?

3:50 p.m.	James Ferris, *Rensselaer Polytechnic Institute*	
4:10 p.m.	Jack Szostak, *Howard Hughes Medical Institute*	
4:30 p.m.	Steven Benner, *University of Florida*	
4:50 p.m.	**Panel 4 Discussion**	Leslie Orgel, Moderator *Salk Institute*
5:50 p.m.	**Concluding Remarks for Panel 4**	Leslie Orgel, Moderator

General Session

6:00 p.m.	**Closing Remarks**	Andrew Knoll, Co-chair *Harvard University* Mary Jane Osborn, Co-chair *University of Connecticut Health Center*
6:10 p.m.	**Adjourn**	

Appendix D

Workshop Participants

Michael Adams, *University of Georgia, Department of Biochemistry and Molecular Biology*
Marc Allen, *NASA Headquarters, Office of Space Science*
Steven Benner, *University of Florida, Department of Chemistry*
David Boal, *Simon Fraser University, Physics Department, Canada*
John P. Bradley, *MVA, Inc.*
Don Button, *University of Alaska at Fairbanks, Institute of Marine Science*
Ron Cowen, *Science News Magazine*
Leonard David, *Space News*
Christian de Duve, *Christian de Duve Institute of Cellular Pathology, Belgium*
Edward F. DeLong, *Monterey Bay Aquarium Research Institute*
Jack Farmer, *Arizona State University, Department of Geology*
James P. Ferris, *Rensselaer Polytechnic Institute, Department of Chemistry*
Dan Fraenkel, *Harvard Medical School, Department of Microbiology and Molecular Genetics*
Phil Harrington, *National Science Foundation*
Olavi Kajander, *University of Kuopio, Department of Biochemistry and Biotechnology, Finland*
Andrew Knoll, *Botanical Museum, Harvard University*
Andrew Lawler, *Science*
Jeffrey Lawrence, *University of Pittsburgh, Department of Biological Sciences*
David McKay, *NASA Johnson Space Center*
Michael Meyer, *NASA Headquarters, Office of Space Science*
Peter Moore, *Yale University, Chemistry Department*
Kenneth Nealson, *Jet Propulsion Laboratory*
Richard Obermann, *U.S. House of Representatives, Committee on Science*
Leslie Orgel, *Salk Institute for Biological Studies*
Mary Jane Osborn, *University of Connecticut Health Center, Department of Microbiology*
Carl Pilcher, *NASA Headquarters, Office of Space Science*
Anthony Reichart, *Nature*

Monica Riley, *Woods Hole Marine Biological Laboratory*
John Rummel, *NASA Headquarters, Office of Space Science*
William Schopf, *University of California at Los Angeles, Department of Earth and Space Sciences*
James Staley, *University of Washington at Seattle, Microbiology Laboratory*
Karl Stetter, *Lehrstuhl für Mikrobiologie, Universität Regensburg, Regensburg, Germany*
Jack Szostak, *Howard Hughes Medical Institute, Department of Molecular Biology*
James Van Etten, *University of Nebraska, Department of Pathology*